Science, Technology, and Society

The Impact
of Science
from
2000 B.C.
to the
18th Century

Science, Technology, and Society

The Impact of Science from 2000 B.C. to the 18th Century

Judson Knight, Neil Schlager, Editors

VOLUME 1

Life Science

CONCORDIA COLLEGE LIBRARY
BRONXVILLE, NY 10708

Detroit • New York • San Diego • San Francisco • Cleveland • New Haven, Conn. • Waterville, Maine • London • Munich

Science, Technology, and Society: The Impact of Science from 2000 B.C. to the 18th Century

Judson Knight, Neil Schlager, Editors

Project Editor
Christine Slovey

Editorial
Carol Nagel, Diane Sawinski

Permissions
Shalice Shah-Caldwell

Imaging and Multimedia
Robert Duncan, Robyn Young

Product Design
Tracey Rowens

Composition
Evi Seoud

Manufacturing
Rita Wimberly

©2002 by U•X•L. U•X•L is an imprint of The Gale Group, Inc., a division of Thomson Learning, Inc.

U•X•L® is a registered trademark used herein under license. Thomson Learning™ is a trademark used herein under license.

For more information, contact
The Gale Group, Inc.
27500 Drake Rd.
Farmington Hills, MI 48331-3535
Or you can visit our Internet site at
http://www.gale.com

ALL RIGHTS RESERVED
No part of this work covered by the copyright hereon may be reproduced or used in any form or by any means—graphic, electronic, or mechanical, including photocopying, recording, taping, Web distribution, or information storage retrieval systems—without the written permission of the publisher.

For permission to use material from this product, submit your request via Web at http://www.gale-edit.com/permissions, or you may download our Permissions Request form and submit your request by fax or mail to:

Permissions Department
The Gale Group, Inc.
27500 Drake Rd.
Farmington Hills, MI 48331-3535

Permissions Hotline:
248-699-8006 or 800-877-4253; ext. 8006
Fax: 248-699-8074 or 800-762-4058

Cover photographs of an Aztec calendar stone and Pythagoras reproduced by permission of the Corbis Corporation (Bellevue).

While every effort has been made to ensure the reliability of the information presented in this publication, The Gale Group, Inc. does not guarantee the accuracy of the data contained herein. The Gale Group, Inc. accepts no payment for listing; and inclusion in the publication of any organization, agency, institution, publication, service, or individual does not imply endorsement of the editors or publisher. Errors brought to the attention of the publisher and verified to the satisfaction of the publisher will be corrected in future editions.

LIBRARY OF CONGRESS CATALOGING-IN-PUBLICATION DATA

Science, technology, and society: the impact of science from 2000 B.C. to the 18th century / Judson Knight, Neil Schlager, editors.

    p. cm.

Includes bibliographical references and index.

    ISBN 0-7876-5653-4 (set : hardcover) — ISBN 0-7876-5654-2 (v. 1) — ISBN 0-7876-5455-0 (v. 2) — 0-7876-5656-9 (v. 3)

    1. Science—History. 2. Science—Social aspects—History. 3. Science and civilization—History. 4. Technology—History. 5. Technology—Social aspects—History. 6. Technology and civilization—History. I. Knight, Judson. II. Schlager, Neil, 1966–

Q125 .S434756 2002   2002005432
509—dc21

Printed in the United States of America
10 9 8 7 6 5 4 3 2 1

Contents

Reader's Guide . . . . *xi*
Advisory Board . . . . *xiii*
Chronology . . . . *xv*
Words to Know . . . . *xxiii*

VOLUME 1

chapter one Life Science

CHAPTER CONTENTS 1
CHRONOLOGY 2
OVERVIEW 3
ESSAYS
Mummification and Medicine in
Ancient Egypt 14
"Alternative" Medicine Begins in the East 19
Greek Physicians Transform the
Life Sciences 29
The Great Doctors of Medieval Islam 36
Fifteen Hundred Years of Superstition 42
The Reawakening of the Life Sciences
in Europe. 54
Printing, Illustration, and Dissection 61
The Medical Discovery of Women 70
Biology and Classification. 79

v

Contents

Three New Worlds 85
The Human Machine 92
The Eighteenth-Century Revolution
in Human Health 99

BIOGRAPHIES
Aristotle. 108
Averroës 110
Avicenna. 112
Galen 113
William Harvey 115
Hippocrates 117
Antoni van Leeuwenhock 118
Carolus Linnaeus 119
Paracelsus. 121
Rhazes 123
Andreas Vesalius 124

BRIEF BIOGRAPHIES 126

RESEARCH AND ACTIVITY IDEAS 139

FOR MORE INFORMATION. 141

INDEX xxxvii

VOLUME 2

chapter two ## Mathematics

CHAPTER CONTENTS. 145

CHRONOLOGY 146

OVERVIEW 147

ESSAYS
The Birth of Mathematics in the Near East 157
Mathematics in China, India, and Beyond 163
The Greeks' New Approach to Mathematics . . . 171
Early Number Systems 178
Hindu-Arabic Numerals and
Mathematical Symbolism 186
Geometry and the Great Unsolved Problems . . 195
Playing with Numbers 208
Patterns and Possibilities. 216

Contents

Angles, Curves, and Surfaces 223

BIOGRAPHIES
Aryabhata 235
René Descartes 236
Eratosthenes 238
Euclid 239
Leonhard Euler 241
Pierre de Fermat 243
Leonardo Fibonacci. 245
al-Khwarizmi. 246
Gottfried Wilhelm von Leibniz. 248
Blaise Pascal 249
Pythagoras. 252

BRIEF BIOGRAPHIES 254

RESEARCH AND ACTIVITY IDEAS 268

FOR MORE INFORMATION 270

chapter three Physical Science

CHAPTER CONTENTS 273

CHRONOLOGY 274

OVERVIEW. 275

ESSAYS
Gazing at the Stars: How Science Began 285
Early Greek Theories of Matter:
What *Is* Everything? 297
Aristotle and Ptolemy:
Wrong Ideas That Defined the World. 308
The Islamic World Takes the Lead 320
Alchemy, Astrology, and Education in
Medieval Europe 331
The Scientific Revolution 342
An Explosion of Knowledge 354

BIOGRAPHIES
Alhazen. 369
Robert Boyle 372
Nicolaus Copernicus 374
Galileo Galilei 376

Contents

Christiaan Huygens 378
Johannes Kepler 380
Antoine-Laurent Lavoisier 382
Isaac Newton 385
Ptolemy 388

BRIEF BIOGRAPHIES 390

RESEARCH AND ACTIVITY IDEAS 404

FOR MORE INFORMATION 406

INDEX xxxvii

VOLUME 3

chapter four Technology and Invention

CHAPTER CONTENTS 409

CHRONOLOGY 410

OVERVIEW 411

ESSAYS
Fire, Metal, and Tools 421
The Roots of Agriculture 430
The Written Word 439
Structures and Cities 447
Infrastructure 459
Machines and Motion 470
Taking to the Water 482
China's Gifts 490
Marking Time 502
The Printed Word 512
Fluids, Pressure, and Heat 523

BIOGRAPHIES
Archimedes 536
Roger Bacon 538
Daniel Bernoulli 540
Benjamin Franklin 542
Johannes Gutenberg 545
Imhotep 547
Leonardo da Vinci 549
Su-sung 551

Contents

Ts'ai Lun 553
James Watt................................. 555
BRIEF BIOGRAPHIES **559**
RESEARCH AND ACTIVITY IDEAS **574**
FOR MORE INFORMATION **577**
INDEX .... **xxxvii**

Reader's Guide

Science, Technology, and Society: The Impact of Science from 2000 B.C. to the 18th Century presents more than eighty topical and biographical essays designed to help students understand the impact that science had on the course of human history between 2000 B.C. and the eighteenth century. Essays examine scientific discoveries and developments and the individuals who made them, showing how social trends and events influenced science and how scientific developments changed people's lives.

Format

Science, Technology, and Society: The Impact of Science from 2000 B.C. to the 18th Century is divided into four chapters across three volumes. The Life Science chapter appears in Volume One. The Mathematics and Physical Science chapters appear in Volume Two. And the Technology and Invention chapter appears in Volume 3. The following sections appear in each chapter:

Chronology: A timeline of key events within the chapter's discipline.

Overview: A summary of the scientific discoveries and developments, trends, and issues within the discipline.

Essays: Topic essays describing major discoveries and developments within the discipline and relating them to social history.

Biographies: Biographical profiles providing personal background on important individuals within the discipline, and often introducing students to additional important issues in science and society during the 3700-year period covered.

Brief Biographies: Brief biographical mentions introducing students to the major accomplishments of other notable scientists, researchers, teachers, and inventors important within the discipline.

Reader's Guide

Research and Activity Ideas: Offering students ideas for reports, presentations, or classroom activities related to the topics discussed in the chapter.

For More Information: Providing sources for further reasearch on the topics and individuals discussed in the chapter.

Other features

Sidebars in every chapter highlight interesting events, issues, or individuals related to the subject. More than 180 black-and-white photographs help illustrate the discoveries and the individuals who made them. In addition, cross-references to subjects discussed in other topic essays are indicated with "see references" in parentheses while cross-references to individuals discussed elsewhere in the title are indicated by boldface type and "see references" in parentheses. Each volume concludes with a cumulative subject index so that students can easily locate the people, places, and events discussed throughout *Science, Technology, and Society: The Impact of Science from 2000 B.C. to the 18th Century*.

Comments and Suggestions

We welcome your comments on *Science, Technology, and Society: The Impact of Science from 2000 B.C. to the 18th Century* and suggestions for other science topics to consider. Please write: Editors, *Science, Technology, and Society: The Impact of Science from 2000 B.C. to the 18th Century*, U•X•L, 27500 Drake Rd., Farmington Hills, Michigan 48331-3535; call toll-free: 1-800-877-4253; fax to (248) 414-5043; or send e-mail via http://www.gale.com.

Advisory Board

Special thanks are due to U•X•L's *Science, Technology, and Society* advisors. The following teachers and media specialists offered invaluable comments and suggestions when this work was in its formative stages:

Dr. Josesph L. Hoffman
Director of Technology
West Bloomfield School District
West Bloomfield, Michigan

Dean Sousanis
Science Chairman
Almont High School
Almont, Michigan

Eric Wisniewski
Math and Science Teacher
Edsel Ford High School
Dearborn, Michigan

Chronology

C. 7000 B.C.E.
First consistent, reliable fire-making technology is developed.

C. 3500 B.C.E.
The Sumerians develop the wheel.

C. 3500–2000 B.C.E.
Early mathematicians of Sumer in Mesopotamia discover the foundations of arithmetic, later developed independently by the Chinese, Indians, and Maya.

C. 2630 B.C.E.
Egyptian architect Imhotep begins building the Step Pyramid, a tomb for the pharaoh Djoser (or Zoser), launching several centuries of pyramid building in Egypt.

C. 1600 B.C.E.
The Phoenicians develop the alphabet.

C. 1500 B.C.E.
Medical practices based on the Vedic religion, a predecessor to Hinduism, appear in India.

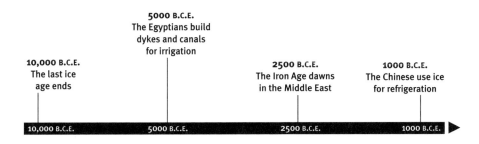

Chronology

C. 600 B.C.E.
Thales of Miletus originates both Western philosophy and the physical sciences by stating that water is the basic substance of the universe.

C. 540 B.C.E.
Persian emperor Cyrus II orders the building of the Royal Road, the first major highway system, which serves as a model for later Roman roads and the interstate highways of today.

C. 530 B.C.E.
Greek philosopher and mathematician Pythagoras establishes a community of followers in a Greek colony on the southern coast of Italy.

C. 400 B.C.E.
Greek physician Hippocrates and his followers establish a medical code of ethics. Hippocrates attributes diseases to natural causes and uses diet and medication to restore the body.

C. 350 B.C.E.
Greek philosopher Aristotle establishes the disciplines of biology and comparative anatomy and makes the first serious attempt to classify animals.

C. 300 B.C.E.
Greek mathematician Euclid writes a geometry textbook entitled *Elements*, which brings together all the known geometric ideas of its time, and is destined to remain the authority on mathematics for some twenty-two hundred years.

C. 300 B.C.E.
The Romans begin developing sophisticated techniques of road building. They also launch the construction of aqueducts capable of carry large quantities of water many miles. Both activities will continue for more than five hundred years.

C. 250 B.C.E.
Greek mathematician and engineer Archimedes invents a number of useful mechanisms, including Archimedes' screw, a device for raising water that is still used in parts of the world today.

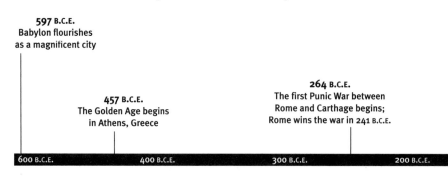

Chronology

C. 140 B.C.E.
Hipparchus, a Greek astronomer, develops the fundamentals of trigonometry.

C. 260 B.C.E.
Aristarchus of Samos, a Greek astronomer, states that the Sun and not Earth is the center of the universe.

C. 350 B.C.E.
Greek philosopher Aristotle proves that Earth is spherical and establishes principles of physics that, though incorrect, will remain influential for the next two thousand years.

C. 425 B.C.E.
Democritus, a Greek philosopher, states that all matter consists of tiny, indivisible particles called atoms.

105 C.E.
Chinese inventor Ts'ai Lun perfects a method for making paper from tree bark, rags, and hemp.

C. 150 C.E.
The Greco-Roman astronomer Ptolemy (Claudius Ptolemaeus) writes his highly influential *Almagest,* which offers a geocentric (Earth-centered) model of the universe that will be widely accepted for the next fifteen hundred years.

C. 160–C. 199 C.E.
Greek physician Galen becomes the most influential figure in medicine and remains so for the next millennium.

C. 400 C.E.
Christian noblewoman Fabiola establishes in Rome (in present-day Italy) the first hospital in western Europe.

400 C.E.
Invaders from central Asia introduce the stirrup to Europe. By allowing a soldier to remain sitting on a horse while delivering a powerful blow, the stirrup makes possible the age of knights and feudalism that follows.

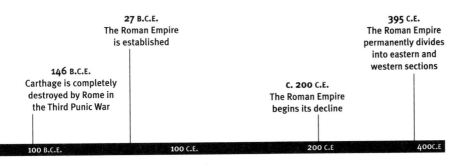

Chronology

499

Aryabhata, a Hindu mathematician and astronomer, writes his *Aryabhatiya*, which describes the Indian numerical system.

c. 600

Block printing makes its first appearance in China.

820

Al-Khwarizmi, an Arab mathematician, writes a mathematical text that introduces the word "algebra" (*al-jabr* in Arabic), as well as Hindu-Arabic numerals, including zero.

c. 800–c. 1300

Medical knowledge flourishes in the Muslim world. Among the distinguished figures of this period are Persian physician known as Rhazes (ar-Razi); Arab physician and philosopher Avicenna (Ibn Sina); and Arab physician Averroës (Ibn Rushd).

c. 1000

The magnetic compass, invented much earlier, is perfected and put into increased use in China.

c. 1000

Arab physicist Alhazen (Abu 'Ali al-Hasan Ibn al-Haytham) argues against the prevailing belief that the eye sends out a light that reflects off of objects.

c. 1000–1170

The medical school of Salerno in Italy, established around 800 C.E., becomes the first institution in Europe to grant medical diplomas. By 1000 it is thriving and brings in students from around the region. In 1170 a professor at the school, Roger of Salerno, publishes the first book on surgery in the West.

1202

In *Liber Abaci*, Italian mathematician Leonardo Fibonacci introduces Hindu-Arabic numerals to the West.

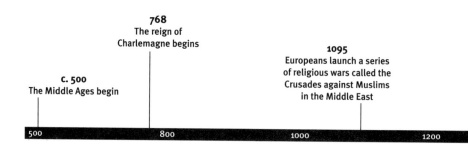

Science, Technology, and Society

Chronology

1224
Holy Roman Emperor Frederick II issues laws regulating the study of medicine, elevating the status of real physicians and diminishing the number of frauds. Later, in 1241, he becomes the first major European ruler to permit dissection of cadavers (cutting open dead bodies for medical study) formerly prohibited by religious law.

1288
The first known guns are made in China. Firearms are first mentioned in Western accounts twenty-five years later, in 1313.

c. 1315
The first mechanical clocks appear in Europe.

c. 1450
German inventor Johannes Gutenberg invents a printing press with movable type, an event that will lead to an explosion of knowledge as new ideas become much easier to disseminate.

1543
In his *De revolutionibus orbium coelestium* (The revolutions of the heavenly spheres), completed in 1530 but not published until the time of his death, Polish astronomer Nicolaus Copernicus proposes a heliocentric, or Sun-centered universe, thus initiating the Scientific Revolution.

1543
Belgian physician Andreas Vesalius publishes his illustrated book of anatomy, *De humani corporis fabrica* (On the structure of the human body), one of the most important works in medical history.

1572
Danish astronomer Tycho Brahe observes a supernova, or exploding star, an event that puts to rest the long-held Aristotelian notion that the heavens are perfect and unchanging.

1587
Italian astronomer Galileo Galilei begins experiments that lead to his law of falling bodies, showing that, contrary to Aristotle, the rate that a body

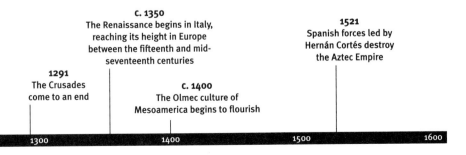

Chronology

falls is independent of its weight, and that all objects will fall at the same rate in a vacuum.

1589
William Lee of England invents the first knitting machine.

1594
Scottish mathematician John Napier develops the idea of logarithms, which will eventually make possible the calculation of difficult multiplication and division problems.

1609
German astronomer Johannes Kepler introduces his laws of planetary motion.

1619
At just twenty-two years of age, French mathematician René Descartes establishes the basics of modern mathematics by applying algebra to geometry and formulating analytic geometry. Eighteen years later, Descartes will publish this breakthrough work in *Discourse on Method*.

1628
English physician William Harvey, considered the founder of modern physiology—the scientific study of the functions, activities, and processes of living things—demonstrates how blood circulates in *Exercitatio anatomica de motu cordis sanguinis in animalius* (On the movement of the heart and blood in animals).

1654
Blaise Pascal sends fellow French mathematician Pierre de Fermat a letter requesting help in solving a problem involving dice and games of chance. This leads to a lively exchange that results in the creation of probability theory.

1661
English physicist and chemist Robert Boyle publishes *The sceptical chymist*, a work regarded by many as the beginning of scientific chemistry.

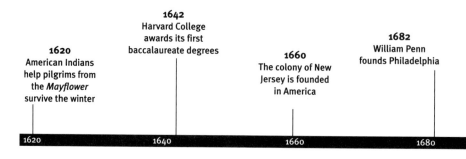

Chronology

1669
British physicist and mathematician Isaac Newton develops a version of calculus, but does not publish his work for many years.

1681
The Languedoc Canal, also known as the Canal du Midi, in southern France is completed. This 150-mile (241-kilometer) waterway will be considered the greatest feat of civil engineering between Roman times and the nineteenth century.

1684
German philosopher and mathematician Gottfried Wilhelm von Leibniz develops his own version of calculus, which he immediately publishes. The years that follow will see a heated debate between supporters of Newton and Leibniz as to who was the true inventor of calculus.

1687
Isaac Newton publishes *Philosophiae naturalis principia mathematica*, generally considered the greatest scientific work ever written, in which he outlines his three laws of motion and offers an equation that becomes the law of universal gravitation.

1735
Swedish botanist Carolus Linnaeus (Carl von Linné) outlines his system for classifying all living things.

1735
German mathematician Leonhard Euler solves the famous Königsberg bridge problem, thus pioneering the areas of graph theory and topology and introducing concepts used today in everything from computer networking to highway design.

1769
Scottish inventor James Watt obtains a patent for his steam engine, which improves on ideas Thomas Newcomen developed half a century earlier.

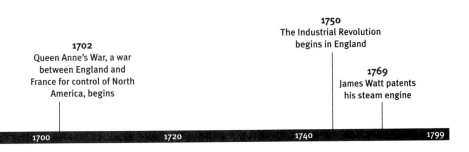

Words to Know

A

abacus An early form of hand-operated calculator that uses movable beads strung along parallel wires inside a frame.

acceleration A change in velocity over time. The acceleration due to gravity, for instance, is 32 feet (9.8 meters) per second per second, meaning that for every second an object falls, its velocity is increasing as well.

acupuncture The insertion of thin needles into specific points of the body in order to relieve pain or treat illness.

alchemy A set of mystical beliefs based on the idea that ordinary matter can be perfected. In the Middle Ages this became a semi-scientific discipline concerned, for instance, with attempts to turn various metals into gold.

algebra A branch of mathematics in which arithmetic operations (for example, addition or multiplication) are generalized. In algebraic equations, symbols represent numbers of unknown value, and the equations themselves are used to find these values.

alternative medicine Medical practices that are not officially recognized by the mainstream medical community.

anatomy The study of the structure of organisms, including the human body.

apothecary One who prepared and sold medicines in medieval times.

applied mathematics The use of mathematics for a specific purpose, as in business or engineering. This is in contrast to pure mathematics.

aqueduct A long pipe, usually mounted on a high stone wall that slopes gently, which is used to carry water from the mountains to the lowlands.

astrolabe A small instrument, used during ancient and medieval times, for calculating the positions of bodies in the heavens.

Words to Know

astronomical clock A clock that, in addition to telling the time, is designed to show the positions of the Sun, Moon, and other celestial bodies.

astrology The study of the positions and the movements of the stars, planets, and other heavenly bodies in the belief that these affect people's lives.

Ayurvedic A term describing a form of medicine, closely tied with the Hindu religion, practiced from ancient times in India.

B

bank In nautical terms, refers to the number of oarsmen along a vertical line. On a trireme, a ship with three tiers, there were three banks. Later ships, however, used more men on a single oar without adding tiers.

bestiary An ancient or medieval catalogue of animal life. Bestiaries were highly unscientific and typically included numerous fictional creatures such as the unicorn.

biology The scientific study of living organisms. Actually a collection of disciplines that includes botany and zoology, biology is (along with medicine) one of the two principal areas of study in the life sciences.

block printing A process whereby a printer carves out the material to be printed on a piece of wood, then inks the wood block and presses it onto paper to create a printed image.

botany A branch of biology concerned with plant life.

C

caduceus A staff with two entwined snakes and two wings at the top, the internationally recognized symbol for medicine.

calculus The branch of mathematics that deals with rates of change and motion.

Cartesian coordinate system A method for identifying points on a plane by assigning to each point a unique set of numbers indicating its location. In the Cartesian system, values of x and y respectively indicate horizontal and vertical distance from the center, designated as (0,0). Cartesian graphs may also be three-dimensional, with a z-axis perpendicular (at a right angle to) both x and y.

case history A record of a particular patient's illness and treatment, often used in medical texts for the purpose of teaching or illustrating a concept.

celestial Of the skies or heavens.

Cesarean section A medical procedure in which a baby is surgically removed from its mother's uterus.

chemistry An area of the physical sciences concerned with the composition, structure, properties, and changes of substances, including elements, compounds, and mixtures.

civil engineering An area of engineering concerned with the design and building of large public works projects such as roads and bridges.

circulatory system The heart, blood, and blood vessels.

circumference The distance around a circle.

civilization A term referring to a society that has all or most of the following: a settled way of life, agriculture, cities, an organized political system, polished tools, and writing.

combinatorics The study of combining objects by various rules to create new arrangements of objects.

conic section The cross-section of a cone when cut by a plane.

constellation A group of stars. In many cases the stars involved are at incredible distances from one another, but they seem, from Earth's perspective, to form groups.

contagion The transmission of a disease by direct or indirect contact.

cosmology A branch of astronomy concerned with the origin, structure, and evolution of the universe.

cosmos The universe.

cube root The cube root of a given number is a number that, when multiplied by itself twice, will produce that number. Thus 2 is the cube root of 8 since $2 \times 2 \times 2 = 8$.

D

decimal fraction A method of representing numbers that are less than 1 by using a decimal point and place value. The number 0.01 is a decimal fraction, whereas 1/100 is a common fraction.

decimal system The number system in use throughout most of the world today, based on 10.

diameter The distance across a circle, as drawn along a path that passes through the center of that circle.

diffraction That which occurs when light rays pass close to an object and bend or separate as a result.

dissection Cutting open dead bodies in order to study their internal characteristics.

domesticate To adapt something (usually a plant or animal) so that it is useful and advantageous for humans.

Words to Know

E

eclipse An event in which one celestial body covers, or otherwise makes it impossible to see, another. In a solar eclipse, the Moon passes between Earth and the Sun, covering the Sun. In a lunar eclipse, Earth comes between the Sun and the full Moon, placing the Moon in Earth's shadow.

ecliptic The great circle of the heavens, which is the apparent path of the Sun as it seems to move across the sky, or the path of Earth as seen from the Sun.

element A substance, made up of only one kind of atom, that cannot be broken down chemically into another substance. Scientists developed this definition around the year 1800; prior to that time, a number of misguided theories prevailed. Most notable among these was the Greek notion of four elements: earth, air, fire, and water.

embryo An unborn animal in the early stages of development. For humans, the embryonic stage is defined as the period of eight weeks following conception, or the fertilization of the egg.

epidemic A disease that affects large numbers of people.

Euclidean geometry Geometry based on the principles laid down by Euclid (c. 330–260 B.C.E.) in his *Elements*, which is concerned primarily with flat, two-dimensional space in which nonparallel lines eventually converge.

F

fluid mechanics The study of fluids (liquids or gases) and their properties. This includes hydrostatics, or the study of fluids at rest, and hydrodynamics or fluid dynamics, the study of fluids in motion.

forceps A medical instrument, shaped like tongs, used to extract a baby during difficult births, as well as for other surgical applications.

G

genus A group of species distinguished by common characteristics.

geocentric Earth-centered.

geology The scientific study of Earth, particularly as revealed through its rocks.

geometry The branch of mathematics concerned with the properties and relationships of points, lines, angles, surfaces, and shapes.

gynecology The branch of medicine concerned with diseases and physical care of women.

H

heliocentric Sun-centered.

herbal A book concerned with plants and their medicinal uses.

herbalist Someone who deals in medicinal herbs. Herbalists played a major role in medieval medicine and continue to do so in alternative medicine today.

Hindu-Arabic numerals The number system in use throughout most of the world today, which uses ten digits, including zero.

horoscope A chart that uses the positions of planets and constellations for the purpose of creating an astrological forecast.

hull The body of a ship.

humors Four fluids (phlegm, blood, yellow bile, and black bile) that, according to Hippocrates and his followers, made up the human body. Imbalances in these humors were supposedly responsible for all illnesses.

hypotenuse The longest side of a right triangle.

hysterectomy An operation in which a woman's uterus is removed.

I

Industrial Revolution A period of rapid development that began in about 1750 and transformed the economies of the West from agriculture-based to manufacturing-based systems.

inertia The tendency of objects in motion to remain in motion, and objects at rest to remain at rest, unless acted upon by some outside force.

infrastructure A system of public works, such as roads and sewers, necessary to the functioning of a society.

inoculation The prevention of a disease by the introduction to the body, in small quantities, of the virus or other microorganism that causes the disease.

inscribe In geometry, to draw a figure inside another one in such a way that it touches, but does not overlap, the boundaries of the larger figure.

intercalation The insertion of an extra month in a year to make the calendar line up with the seasons.

irrational number A number that cannot be expressed as the ratio of any two whole numbers because in decimal form, irrational numbers such as pi (π) neither terminate (come to an end) nor fall into a repeating pattern.

K

kinematics The study of how objects move.

L

life expectancy A calculation, based on statistical data, of the average life span of an organism.

lock A device on a canal that allows a vessel to negotiate changes in altitude by raising or lowering the water level.

logarithm The power to which a given number, called a base, must be raised to yield a given product, for instance, $10^2 = 100$. In logarithmic terms this would be expressed thus: $\log_{10} 100 = 2$.

logic A system of reasoning for reaching valid conclusions about concepts, and for assessing the validity of a conclusion that has been reached.

lunar calendar A measure of the year based on twelve lunar months (the time it takes the Moon to revolve around Earth), which lasts about 354.37 days.

luni-solar calendar A combination of lunar and solar calendars, which uses intercalation to align the two.

M

machine Any device that transmits, modifies, or magnifies force for a specific purpose. Machines typically either alter the amount of force applied, or the direction along which it is applied. A simple machine is a machine with only one or two parts.

magic square A square set of boxes—the number must always be a square number such as 9, 16, or 25—containing different numbers. In a magic square, the sum of the numbers in each row, column, or diagonal is the same.

matter A physical substance that occupies space, possesses mass, and is ultimately convertible to energy. Matter can be a gas, liquid, or solid.

mechanics A branch of physics concerned with the study of bodies in motion.

medieval Having to do with the Middle Ages (c. 500–c. 1500 C.E.).

menstruation The monthly discharge of blood and other materials from the uterus of a nonpregnant female primate (a class of animals including apes as well as humans) of breeding age.

metallurgy The science and technology of metals.

midwife A person who assists women in childbirth. In premodern times, midwives were almost always female and seldom had formal medical training.

movable-type printing A process in which a printer uses precast pieces of type, representing letters and other symbols, which are inserted into a frame and used to print a document.

N

negative number A number smaller than 0.

Neolithic Revolution A term describing a series of changes that occurred between about twenty thousand and six thousand years ago, when humans began to produce much more sophisticated, polished tools, as well as pottery and woven materials.

non-Euclidean geometry A form of geometry, developed in the 1800s, that is concerned primarily with curved (and sometimes three-dimensional) space in which it is possible that nonparallel lines do not converge.

notation Symbols to represent numbers or operations such as addition and subtraction.

number theory The study of the properties of numbers, and the relationships between them.

numerology The belief that numbers have special meanings of a spiritual nature, and can be used to predict the future.

O

obstetrics The branch of medicine concerned with childbirth.

occupational health An area of medicine concerned with the health hazards related to specific professions.

optics The study of light and vision.

ovary The part of the female anatomy in which eggs (which can be fertilized by sperm to create offspring) are produced.

P

periodical A publication, such as a newspaper or magazine, that comes out on a periodic basis, for example, every day or every month.

pharmacology A branch of medical science concerned with medicines and drugs.

philosophy The area of study that seeks to provide a general understanding of reality.

phonogram A written symbol that represents a specific syllable.

Words to Know

physics An area of the physical sciences that is concerned with matter, energy, and the interactions between them.

physiology The scientific study of the functions, activities, and processes of living things.

pi The ratio between the circumference of (distance around) a circle and its diameter (distance across). Pronounced "pie" and represented by the Greek letter π, pi has a value of 3.14159265+. This is its value to eight decimal places; in fact pi is an irrational number.

pictogram A written symbol that looks like the thing it represents.

place value The use of a number's position to indicate its size. In a number such as 1239, for instance, the 2 does not stand for 2, but for 200. Whereas the Hindu-Arabic numeral system in use today has a built-in concept of place value, the Roman numeral system used during the Middle Ages (c. 500–c. 1500) did not.

plane method The technique of performing geometric work using only a compass and an unmarked straightedge.

polygon A closed shape with three or more sides, all straight.

postulate A basic principle or established rule. Sometimes called an axiom.

prehistoric Anything that existed before the development of written language.

prime number A number that can be divided only by itself and 1.

probability theory A branch of mathematics devoted to predicting the likelihood that a particular event will occur. Such predictions are typically based on statistical data.

prognosis A prediction regarding the course and outcome of a disease based upon previous observation of similar cases.

proof A step-by-step process of proving certain ideas in geometry by referring to already established propositions.

public health A set of policies and methods for protecting and improving the health of a community through efforts that include disease prevention, health education, and sanitation.

pulmonary circulation The movement of blood from the right ventricle of the heart into the lungs via the pulmonary artery, and from the lungs to the left ventricle through the pulmonary veins.

pure mathematics Mathematics for its own sake, rather than for a specific application. Compare with applied mathematics.

pythagorean theorem A rule, attributed to the Greek mathematician Pythagoras (c. 580–c. 500 B.C.E.) but probably derived much earlier, which states that for every right triangle, the square of the hypotenuse is equal to the sum of the squares of the other two sides.

Q

quack Someone who falsely claims to possess medical skill and the ability to provide proper treatment.

quarantine The isolation or separation of people or items in order to prevent the spread of communicable diseases.

R

ratio The relationship between two numbers or values. All fractions are ratios, but some ratios such as π (pi, the relationship between the circumference and diameter of a circle) cannot be expressed as fractions.

reflection That which occurs when light rays strike a smooth surface and bounce off at an angle equal to that of the incoming rays.

Reformation A religious movement in the 1500s that ultimately led to the rejection of Roman Catholicism by various groups and the formation of Protestant religious denominations.

refraction The bending of light as it passes at an angle from one transparent material into a second transparent material. Refraction accounts for the fact that objects under water appear to have a different size and location than they have in air.

retrograde motion The apparent backward movement, or reversal of direction, by outer planets in the solar system. In fact retrograde motion is simply an optical illusion, created by the fact that Earth is orbiting the Sun much faster than the outer planets are.

right triangle A triangle with one right angle, or 90° angle. (Since the total measurement of the three angles in a triangle is 180°, no triangle can have more than one right angle.)

S

scientific method A set of principles and procedures for systematic study, introduced primarily by Galileo Galilei, and still used in the sciences. The scientific method consists of four essential parts: the statement of a problem to be studied; the gathering of scientific data through observation and experimentation; the formulation of hypotheses or theories; and the testing of those hypotheses. The results of testing may lead to a restatement of the problem, or an entirely new problem to be analyzed, which starts the process over again.

Scientific Revolution A period of accelerated scientific discovery that completely reshaped the world. Usually dated from about 1550 to 1700, the Scientific Revolution saw the origination of the scientific method and the introduction of ideas such as the heliocentric universe and gravity. Its leading figures included Nicolaus Copernicus, Galileo Galilei, and Isaac Newton.

scribe A member of a small and very powerful group in ancient society who knew how to read and write.

scurvy An illness caused by a lack of vitamin C that results in swollen joints, bleeding gums, loose teeth, and an inability to recover from wounds.

sexagesimal system A number system based on 60.

smallpox A viral infection accompanied by fevers and chills, and characterized by the formation of a rash over large parts of the body. As the effects of the illness continue, the rash turns to pus-filled bumps or papules that, when infected, can cause death. Even those who survive, however, bear the scars left by the eruption of the papules.

solar calendar A measure of the year based on Earth's revolution around the Sun, which takes 365.2422 days. This is the calendar used in most of the world today.

species A category of closely related organisms. A species is usually defined by the ability of its members to breed with one another and by their inability to breed with members of other species.

square The square of a number is that number multiplied by itself. Thus 25, for instance, is the square of 5, since $5 \times 5 = 25$.

square root The square root of a given number is a number that, when multiplied by itself, will produce that number. Thus 2 is the square root of 4 since $2 \times 2 = 4$.

standardize To establish a common standard with which measurements, or measuring devices, must be consistent.

star catalogue A listing of the known stars with their names, positions, and movements.

statistics A branch of mathematics concerned with the collection and analysis of numerical data.

T

technology The application of knowledge to make the performance of physical tasks easier.

terrestrial Of the Earth.

theorem A proposition based on one or more postulates.

thermodynamics The study of the relationships between heat, energy, and work. (Work, closely related to the concept of power, is a term in physics that has a definition different from its everyday meaning.)

tier A level. In discussing ancient ships, the levels of rowers, one atop the other, powering the craft.

tool A handheld device that aids in the accomplishment of tasks.

trigonometry The study of the properties of triangles—in particular, the relationships between the various sides of a right triangle—as well as the properties of points on a circle.

U

uterus An organ in the body of a female mammal that holds offspring during their stages of development prior to birth.

V

vacuum An area devoid of matter, even air.

vagina A canal leading from the exterior of a woman's body to the uterus.

ventricle A chamber in the heart.

velocity Speed in a certain direction.

vernacular The language of the people. During the Middle Ages, when educated people communicated with each other in Latin, "vernacular" referred to everyday national languages such as Italian, English, German, or French.

W

watershed A ridge of high land that divides areas drained by different river systems.

whole number A number, such as 1, or 23, or 2,765,014, that includes no fractions.

Z

zodiac An imaginary band in the heavens, divided into twelve constellations or astrological signs.

zoology A branch of biology concerned with animal life.

Science, Technology, and Society

The Impact
of Science
from
2000 B.C.
to the
18th Century

chapter one Life Science

Chronology 2
Overview 3
Essays 14
Biographies 108
Brief Biographies 126
Research and Activity Ideas 139
For More Information 141

CHRONOLOGY

c. 1500 B.C.E.
Medical practices based on the Vedic religion, a predecessor to Hinduism, appear in India.

c. 400 B.C.E.
Greek physician Hippocrates and his followers establish a medical code of ethics. Hippocrates attributes diseases to natural causes and uses diet and medication to restore the body.

c. 350 B.C.E.
Aristotle establishes the disciplines of biology and comparative anatomy and makes the first serious attempt to classify animals.

c. 160–c. 199 C.E.
Greek physician Galen becomes the most influential figure in medicine and remains so for the next millennium.

c. 400
Christian noblewoman Fabiola establishes in Rome (in present-day Italy) the first hospital in western Europe.

c. 800–c. 1300
Medical knowledge flourishes in the Muslim world. Among the distinguished figures of this period are Persian physician known as Rhazes, or ar-Razi; Arab physician and philosopher Avicenna, or Ibn Sina; and Arab physician Averroës, or Ibn Rushd.

c. 1000–1170
The medical school of Salerno in Italy, established around 800 C.E., becomes the first institution in Europe to grant medical diplomas. By 1000 it is thriving and brings in students from around the region. In 1170 a professor at the school, Roger of Salerno, publishes the first book on surgery in the West.

1224
Holy Roman Emperor Frederick II issues laws regulating the study of medicine, elevating the status of real physicians and diminishing the number of frauds. Later, in 1241, he becomes the first major European ruler to permit dissection of cadavers (cutting open dead bodies for medical study), formerly prohibited by religious law.

1347–51
A fatal disease known as the Black Death, or bubonic plague, sweeps Europe, killing approximately thirty million people.

1543
Andreas Vesalius publishes his illustrated book of anatomy, *De humani corporis fabrica* (On the structure of the human body), one of the most important works in medical history.

1628
English physician William Harvey, considered the founder of modern physiology—the scientific study of the functions, activities, and processes of living things—demonstrates how blood circulates in *Exercitatio anatomica de motu cordis sanguinis in animalius* (On the movement of the heart and blood in animals).

1735
Swedish botanist Carolus Linnaeus (Carl von Linné) outlines his system for classifying all living things by using two names, one for genus and one for species.

OVERVIEW

Chapter One

LIFE SCIENCE

Little is written of the life sciences prior to about 500 B.C.E., though cave paintings dating back some thirty thousand years provide evidence that early artists knew a great deal about animal anatomy. Prehistoric remains found at archaeological sites indicate the use of primitive forms of surgery—even brain surgery.

Egyptian methods of preparing mummies for burial dating back as far as 3500 B.C.E. suggest a relatively sophisticated understanding of internal human anatomy. Over four thousand years ago the Chinese developed acupuncture, involving the use of needles inserted in the body to treat illness, and medical study flourished in India with the rise of Ayurvedic medicine in about 1500 B.C.E. During the centuries that followed, Indian doctors introduced a number of advanced techniques such as inoculation and plastic surgery.

The life sciences in Greece and Rome

The formal study of the life sciences actually began during the fourth century B.C.E., a period known as the "golden age," in which Greek civilization was at its peak. One of the most important figures of this era was **Hippocrates** (c. 460–c. 377 B.C.E.; see biography in this chapter), often referred to as the "father of medicine." Hippocrates may be best known for the Hippocratic oath, a code of medical ethics for physicians. Yet the oath, probably composed by his followers rather than Hippocrates himself, was just one of his many contributions to the field.

Perhaps his most important legacy was his emphasis on the physical (rather than the spiritual) causes of disease. Before his time, when a person became sick, it was often believed that he or she had angered one of the gods (as illustrated in Greek mythology). However, Hippocrates taught that illness resulted from imbalances in the body, rather than punishment for bad behavior.

On the other hand, Hippocrates also subscribed to the highly unscientific humoral theory, based on the belief that the body was made up of four humors, or fluids, which supposedly governed human health and behavior.

Life Science

OVERVIEW

The influence of humoral theory, introduced by the Greek philosopher Empedocles (c. 490–430 B.C.E.), would continue for almost two thousand years, during which time it spawned numerous misguided medical treatments such as removing "excess" blood from patients. Yet most educated figures of ancient times believed in humoral theory, among them **Galen** (129–c. 199; see biography in this chapter), a prominent Greek physician of the Roman Empire, who believed that all diseases were the result of irregular or improper distribution of the four humors. Just as Hippocrates's career marked the beginning of Greek progress in the study of life sciences, Galen's career marked the end.

Aristotle, botany, and biology
The body of work left behind by Hippocrates and Galen contained both truth and error, but it would greatly influence Western thinking until the Renaissance began after about 1350. The Renaissance was the period of artistic and intellectual rebirth that marked the end of the Middle Ages.

The same was true of Greek philosopher **Aristotle** (384–322 B.C.E.; see biography in this chapter), one of the most influential thinkers who ever lived. Aristotle addressed virtually every aspect of human study and endeavor, some more successfully than others. For example, his ideas on physics proved to be highly misguided (see essay "Aristotle and Ptolemy: Wrong Ideas That Defined the World" in the Physical Science chapter), but in the areas of biology and botany he exerted a highly positive influence.

Indeed, Aristotle may be considered the "father of biology," just as his student Theophrastus (c. 372–c. 287 B.C.E.) is the "father of botany," the study of plants. Central to the work of Aristotle and his followers was their emphasis on classification. Aristotle made the first serious attempt to organize animal species with related characteristics, and he developed theories on reproduction and the inheritance of various characteristics.

The Roman postscript to Greece
Rome's contribution to scientific knowledge was relatively minor compared to that of Greece. In general, Roman civilization modeled itself on the Greeks, but what Rome lacked in cultural genius it made up for in military strength. During a period from about 200 B.C.E. to about 200 C.E., Rome conquered a huge territory stretching from Britain to the Middle East, and from north Africa to western Russia. To support its vast operations, it developed a system of military medical facilities that became the model for the first hospitals in the West.

As Roman civilization began to decline after 200 C.E., however, the pace of learning in the West came to a virtual standstill. One of the only vital forces in the Roman Empire was Christianity. In the centuries that followed,

Life Science

OVERVIEW

Greek philosopher Aristotle made the first attempt to organize animal species with related characteristics. He was the first to identify dolphins as mammals. (© Stephen Frink. Reproduced by permission of the Corbis Corporation.)

known as the Middle Ages (roughly 500 C.E. to 1500 C.E.), the church would largely assume control of education and society in general. Christians also established the first significant charity hospitals, which sprang up in locations from Britain to what is now Turkey, after about 400 C.E.

Medieval science

The Roman Empire had been divided in half by Roman emperor Constantine in 313. The Western Roman Empire had collapsed by about 476, but the eastern half, known as the Byzantine Empire, would continue to exist until 1453. Yet for the most part the Byzantines contributed little to the advancement of the sciences. As for western Europe, it was plunged into a state of disorder that continued throughout the first half of the Middle Ages.

During this time, European doctors and scientists were not conducting significant new research, and few dared to challenge the ideas that had been around for centuries. In fact, western Europe had been so completely devastated by invasions from outside forces, such as the Huns (a nomadic people from central Asia who seized control of large portions of central and eastern Europe under its leader Attila in about 450 C.E.), that most scientific texts from ancient times had been destroyed.

Superstition often took the place of science in the Middle Ages, and a person took his chances when seeking treatment from one of the empirics or alternative practitioners who passed for doctors in medieval times (the

Life Science

OVERVIEW

*OPPOSITE PAGE
The Crusades
(1095–1291) brought
Europeans into contact
with the advanced
scientific learning of
the East.
(Reproduced by
permission of the Corbis
Corporation.)*

Middle Ages). By about 1000, people were seeking relief from religious miracle healers, or making pilgrimages to Jerusalem and other shrines in the Holy Land (the area equivalent to modern-day Israel and Lebanon) in the hope that God would cure their ills. Those pilgrimages in turn set the stage for one of the most important events in European history: the Crusades (1095–1291).

The Holy Land, along with much of the Middle East, had fallen under the control of Muslims. In 1095 the pope, as leader of the Catholic Church and unofficial leader of western Europe, called for a holy war to place the Holy Land under Christian control. The reason, supposedly, was that Muslims had harassed pilgrims to the Holy Land; but in fact many historians agree that the Crusades were motivated more by a desire for wealth and land than by any noble purposes. Militarily, the Crusades were a failure. Yet because they placed the West in contact with a far more advanced civilization, the Arab world, these wars proved the salvation of Europe.

Scientific progress in the Arab world

The Muslim world had experienced one of history's great periods of cultural and intellectual awakening in the centuries between 800 and 1200. Among the distinguished Muslim physicians of that era were **Rhazes** (ar-Razi; c. 865–c. 923; see biography in this chapter), who distinguished between measles and smallpox, and the physician-philosophers **Avicenna** (Ibn Sina; 980–1037; see biography in this chapter) and **Averroës** (Ibn Rushd; 1126–1198; see biography in this chapter), each of whom produced numerous works on medicine and science that remained in use for hundreds of years.

Part of what fueled the explosion of knowledge in the Muslim world was the rediscovery of writings by Galen, Hippocrates, and Aristotle. Ironically, then, it was the Arabs and Persians (Iranians) of the Middle East who introduced the West to the knowledge of the Greeks; Greek texts that made their way into western Europe after the Crusades had to be translated from Arabic. In many cases, the Greek originals had been lost, and without the Arabic versions made later, those classics would have been lost as well.

The reawakening of European science

The Crusades spelled the beginning of the end of medieval times: thanks in large part to the knowledge they gained in the Middle East, Europeans were poised for a reawakening. Although the first universities did not make their appearance in Europe until about 1200, the first medical school had been founded at Salerno, Italy, in the 800s, and in about 1170

Life Science

OVERVIEW

one of its professors produced the first practical guide to surgery. The success of the Salerno school led to the establishment of medical schools in Paris, France, and other cities throughout Europe.

At that time, much of central Europe belonged to a loose confederation of principalities known as the Holy Roman Empire. Despite its magnificent name, the empire had limited political power, and what authority it had depended on the personal abilities of its rulers. One of the most powerful of these was Frederick II (1194–1250), who in 1224 issued the first laws regulating the study of medicine, thus helping to ensure that only qualified physicians, and not frauds, earned the right to practice as doctors. In 1241 he became the first major European ruler to permit dissection of cadavers (that is, cutting open dead bodies for medical study), formerly prohibited by religious law.

Late medieval botany and zoology

Frederick also wrote a book on falconry, the art of hunting with birds of prey, and thus added to the growing knowledge of zoology and botany, the study of animals and plants, respectively. Due to the medicinal qualities of certain herbs, botany received much more attention throughout the Middle Ages, though as with other scientific disciplines, most knowledge was hundreds of years old. As for zoology, books called bestiaries discussed exotic animals, but many of these were based on erroneous descriptions brought back to Europe by travelers. Some animals, such as the unicorn, were entirely fictional.

By the 1100s scientists had begun studying plants not only for their medicinal uses, but also for their structures and relationships to other organisms. One of the most important figures of this period was Albertus Magnus (c. 1200–1280), who, though greatly influenced by Aristotle's work on biology, also made many original observations. Like most medieval scholars, Albertus Magnus was a priest in the Catholic Church, yet he helped convince church authorities that the study of nature did not imply a questioning of God or God's existence.

Confronting diseases

By the late Middle Ages, a number of hospitals, many of them modeled on the medical facilities of the Muslim world, had sprung up throughout Europe. Maintained by the church and operated by monks and nuns, these institutions provided medical care for the poor. But conditions were often dirty, and the state of treatment left much to be desired. For example, the first important mental institution, Saint Mary of Bethlehem, was established in London, England, in 1247. Conditions there were so miserable that an abbreviated form of its name, "bedlam," came to mean a state of chaotic disorder.

Life Science

OVERVIEW

An engraving from William Hogarth's A Rake's Progress *depicts a scene in "Bedlam," the St. Mary of Bethlehem mental institution.* (Reproduced by permission of the Corbis Corporation.)

Whereas mental illness affected relatively few people, Europe was about to confront a disease that would ultimately kill some thirty million people: the Black Death (1347–51). Plagues, or epidemics, had been a problem since ancient times, as had contagious diseases such as the skin ailment known as leprosy. Nothing, however, approached the scale of the Black Death, or bubonic plague, sometimes known simply as the plague.

It would be many centuries before scientists recognized the role played by rats, fleas, and bacteria in transmitting the plague, but many noticed that improved public sanitation did seem to slow its spread. Indeed, it is now clear that many of the diseases that swept the premodern world were spread by the dumping of garbage and human waste on the streets of Europe's major cities. The first monarch to attempt a reversal of this disastrous practice was Richard II of England (1367–1400), who in 1388 issued the first sanitary laws.

The birth of modern science

If any single event can be said to have marked the beginning of the modern era, it was the development of the printing press in 1450. This invention enabled the wide distribution of illustrated anatomy books such as *De humani corporis fabrica,* a groundbreaking work published in 1543 by

Life Science

OVERVIEW

An alchemist (seated left wearing eyeglasses) and his assistants in the alchemist's shop. (Reproduced by permission of The Granger Collection, New York.)

Andreas Vesalius (1514–1564; see biography in this chapter). Later, books called herbals, which gave information on the medicinal uses of plants, were published, as were many more illustrated works of anatomy. Printing also made possible wider distribution of ancient texts, and this led to a questioning of old ideas and a search for better information.

From about 1450 to 1600, the life sciences remained a mixture of old practices and new. The Swiss scientist **Paracelsus** (1493–1541; see biography in this chapter) embodied this amalgam. While he condemned medieval forms of medical treatment as barbaric, he also subscribed to the medieval chemical pseudoscience known as alchemy. Yet alchemy, which involved ideas such as the transformation of ordinary metals into gold, paved the way for genuine chemistry. Likewise, Paracelsus's interest in alchemy led to an emphasis on man-made compounds versus plant-derived remedies, a step toward the development of modern pharmaceuticals.

Diseases such as the bubonic plague, yellow fever, and leprosy abounded throughout those years. The end of the 1400s brought the first major outbreak of a sexually transmitted disease: syphilis. Despite progress in medical training, many doctors in Europe, designated as barber-surgeons, possessed minimal qualifications. These practitioners used their razors to lance (cut open) boils, extract blood, and perform other medical duties in addition to the tasks normally associated with a barber.

Still, there were signs of change on the horizon. The French surgeon Ambroise Paré (1510–1590) introduced the practice of ligature, or tying off severed blood vessels, and condemned the use of boiling oil as a treatment for gunshot wounds. Italian physicians Bartolommeo Eustachio (1520–1574) and Gabriele Falloppio (1523–1562), whose names are today commemorated in those of body parts (Eustachian tube, fallopian tube), conducted important studies of the inner ear and female reproductive systems, respectively.

Life Science

OVERVIEW

The discovery of new worlds
Indeed, doctors seem to have "discovered" women only during the early modern era. Up until then, midwives trained in traditional medicine delivered babies. Though these midwives were unsung heroes, and their treatment of patients was undoubtedly better than that offered by many university-trained physicians of the time, the incorporation of childbirth into medical study was an important step in the development of the life sciences.

Everywhere, the old rules were changing. French philosopher and natural scientist **René Descartes** (1596–1650; see biography in Math chapter) developed an idea that was considered innovative at the time—that the human body was like a machine or mechanism, which meant its functions were not a mystery known only to God. He maintained that the body could be analyzed just as one might study the workings of an incredibly intricate watch. Descartes's contemporary **William Harvey** (1578–1657; see biography in this chapter) illustrated this principle by accurately describing the human circulatory system, preparing the way for the first blood transfusions in the 1660s.

Much of the progress made in the life sciences from about 1450 to 1700 resulted from, quite literally, the discovery of new worlds. The conquest of the New World, or the Americas, introduced Europeans to all manner of new plants and animals, including corn, potatoes, tomatoes, turkeys, and tobacco. Europeans also brought new species to the New World. Some, such as the horse, would benefit the native inhabitants, though at first horses helped the Europeans conquer the Indians. Others, such as disease-carrying rats, often arrived on ships. Diseases from Europe such as smallpox wiped out native populations.

European explorers also discovered new species in other lands throughout Asia, Africa, and Australia. But perhaps the greatest discoveries of that era were those aided by the microscope, invented in the Netherlands between 1590 and 1608. For the first time, scientists glimpsed the vastly complex world beyond the view of the naked eye. This invention led to important discoveries by the Dutch naturalists **Antoni van Leeuwenhoek** (1632–1723; see biography in this chapter)

Life Science

OVERVIEW

and Jan Swammerdam (1637–1680), as well as a number of botanists and zoologists.

The revolution in the eighteenth century

From 1700 to 1799 more advances in the life sciences took place than had occurred in the entire history of civilization up to that time. It is perhaps surprising, then, that this period is not characterized by the same sorts of towering figures, such as Aristotle, Rhazes, Paracelsus, or William Harvey, whose work dominated earlier ages.

Unlike the myriad of subjects studied by the generalists of past centuries, the work of eighteenth-century life scientists had become increasingly specialized. A typical figure of the era was the French dentist Pierre Fauchard (1678–1761), who made advances in techniques for filling cavities. Fauchard's work may not have been very glamorous, but it helped improve the general quality of life.

A survey of eighteenth-century progress

Among the advances in anatomy and physiology that took place during the eighteenth century was the discovery that the body's organs are made up of many different kinds of tissues, an understanding that opened the way for later studies on the cellular organization of tissues. Another important development was the effort to discover relationships between anatomy and disease, which led to the foundation of pathology (the study of the causes of disease) as a branch of the life sciences.

An American Indian offers tobacco to a European explorer. Tobacco was among the many new plant species that Europeans were introduced to in the Americas. (Reproduced by permission of the Corbis Corporation.)

Life Science

OVERVIEW

Studies in plant physiology brought an understanding of photosynthesis (the process by which green plants absorb carbon dioxide and release oxygen in the presence of sunlight), pollination, and the means by which water moves through plant tissues. Particularly fruitful were eighteenth-century studies in animal physiology, which helped explain how the brain communicates with the body. Scientists of the era demonstrated that nerves stimulate muscle contraction, and that nerves themselves are in turn governed by electrical impulses.

A number of advances in the area of disease prevention also took place in the eighteenth century. The discovery of high cancer rates among chimney sweeps, for instance, marked the first established link between environmental factors and cancer, a topic still under heavy investigation by researchers today. Likewise, the experimental demonstration that citrus fruit could prevent and cure scurvy, a bone disease common among sailors on long voyages, would later lead to the discovery of vitamin C and of vitamins in general. The latter part of the century saw development of the first modern vaccines against smallpox, as well as the birth of modern concepts concerning public health.

Obscure eighteenth-century physicians and life scientists laid the groundwork for the findings of famous nineteenth century figures, such as English naturalist Charles Darwin (1809–1882) and French chemist and microbiologist Louis Pasteur (1822–1895). Yet one figure of the eighteenth century stands out from the rest, though not for a scientific discovery or medical technique; instead, Swedish botanist **Carolus Linnaeus** (1707–1778; see biography in this chapter), or Carlos von Linné as he is known in the West, developed a method of classifying plants that would eventually extend to animals as well. Linnaeus's binomial, or two-name system, which identifies genus and species, greatly simplified the study of the life sciences and made possible many of the advances that followed.

Making sense of it all

As Linnaeus's career showed, one of the principal challenges facing the life sciences in the eighteenth century was the need to make sense out of the accumulating mass of detail concerning the natural world. Scientists were focused on how things in the natural world related to each other, to humanity, and to the concept of God.

For instance, Descartes offered a mechanistic view of life and maintained that it was governed by the same sort of impersonal principles that **Isaac Newton** (1642–1727; see biography in Physical Science chapter) had discovered in the physical sciences. (Isaac Newton published

Life Science

ESSAYS

Philosophiae naturalis principia mathematica in 1687, generally considered the greatest scientific work ever written, in which he outlines his three laws of motion and offers an equation that became the law of universal gravitation.) (See essay "The Scientific Revolution" in the Physical Science chapter.) Others, however, saw life processes as involving principles unique to life and different from those in the nonliving world. They held, for instance, that life can only be explained as the result of deliberate actions on the part of a creator—God.

Another debate concerned the question of whether an individual is preformed in the womb, developing from a tiny model of a human body, or only gradually develops through growth. Today, of course, it is clear that the an individual gradually develops through a series of processes, known as *epigenesis*. Likewise, scientists have long since resolved an issue that plagued the life sciences in the eighteenth century: spontaneous generation—the idea that living organisms can arise from nonliving matter. According to this principle, if food is left in an empty room and roaches appear, this is because the food has spontaneously (instantly) generated the roaches.

Descartes's ideas pose an interesting quandary. Of course, the human body is a mechanism, and the forces that govern its operation do appear to be impersonal. And, just as Descartes held, there is a sharp dividing line between living and nonliving things, as nineteenth-century discoveries in organic chemistry would show. But none of these facts contradict the idea that there might be some form of intelligent design underlying the intricate order in all living things.

Twenty-first-century studies in mathematics show, for instance, that it would be impossible to produce even a bacterium by a series of merely random biochemical processes. Although this does not prove the existence of God, the questions of design, randomness, and order, when viewed from a scientific perspective, continue to offer promising ground for speculation and study.

ESSAYS □

□ MUMMIFICATION AND MEDICINE IN ANCIENT EGYPT

The countries that make up North America and western Europe, as well as European-dominated nations elsewhere, are known collectively as "the

Life Science

ESSAYS

Egyptian embalmers prepare a body for mummification. Embalmers were among the most respected and admired members of the community. (©Bettmann/Corbis. Reproduced by permission of the Corbis Corporation.)

West." In their language, political structures, and cultural institutions, these lands all bear the stamp of ancient Rome, which modeled its civilization on that of Greece—which in turn was influenced by Egypt. Thus it can be said that the ultimate source of Western civilization was Egypt.

Though Egypt was established as a unified kingdom in about 3000 B.C.E., people had been living along the fertile river valley of the Nile in north Africa for thousands of years. The unification of Egypt led to the establishment of the world's first powerful state, one that remained politically significant for the next thousand years. That power is symbolized in the pyramids, vast tombs built by Egypt's kings, or pharaohs. The Great Pyramid of Giza, just across the Nile from Cairo, Egypt's modern capital, remains one of the most impressive structures in the world—and until just a few hundred years ago, it was also the tallest.

Why a pharaoh would go to such effort (and in particular, demand such sacrifice from his people) for a tomb lies in the fact that to the Egyptians, the pyramids were not merely tombs: they provided a house in which the deceased ruler and his family would live during the afterlife. In

Life Science

ESSAYS

ancient Egypt people believed that life after death was much more important than the here and now; therefore, they spent much of their time preparing for the afterlife. Because of these strong beliefs, it became necessary to develop means of preserving the body. Out of this need grew some of the first serious studies in the life sciences.

The embalmer's work
Though their job required a basic understanding of anatomy, embalmers were in a class distinct from physicians. In modern America, people tend to regard the preparation of bodies for burial with distaste, but this was not so for the Egyptians, who described the funeral parlors of their day by a term meaning "the beautiful house." The embalmer was one of the most respected and admired members of the community, occupying a place akin to that of a famous nuclear scientist or astronomer today.

Since the Egyptians believed that life after death was possible only if the body could be preserved, it became enormously important to protect the body from decay. To stop this decay, embalmers dried the body by an elaborate process. First they removed all internal organs (except for the heart, which they considered the seat of intelligence). Then they pulled the brain out through the nose with iron hooks. These organs and brain were placed in jars and sealed with the body inside the tomb.

Next, the embalmer treated the body with natron, a drying agent collected from shallow ponds near the Nile. The embalmers would pack the corpse's internal cavities with bags of natron, cover the body with more of the drying agent, and leave it for a period of forty days. Finally, the embalmers packed mud, sand, or linen under the skin to fill it out, and then wrapped the body in layers of linen. They poured heated resins, or tree saps, over each layer. When these resins hardened, they provided a tough casing that preserved the body within.

There is a simple reason why only the pharaohs' mummies and tombs have survived to the present day: money. Not only did the building of pyramids require considerable wealth and power, but the services of the embalmer were also expensive. So, too, were the materials used: none of the resins came from Egypt itself, making mummification a very costly process.

High and lows of Egyptian medicine
Like the resins, many of the over seven hundred drugs applied by Egyptian physicians came from other places. Evidence indicates that the Egyptians imported medicines from civilizations on the Mediterranean Sea, from other regions of Africa, and from Asia—perhaps even from as far

Dietary Laws of Ancient Israel

A kosher meat market in New York.
(Reproduced by permission of AP/Wide World Photos.)

The first five books of the Bible, referred to as the Torah or *Pentateuch,* which provide the foundation of Jewish faith and law, include a set of dietary restrictions, referred to as *kashruth,* or kashrut (kahsh-ROOT). The most well-known dietary restriction is the prohibition against eating pork. Not all Jews today still adhere to these laws, and scholars have long debated their religious significance. Nonetheless, it is clear that the dietary restrictions of the Torah presented a good, practical set of guidelines for preserving health many thousands of years before scientists recognized bacteria and other causes of food spoilage.

Pork, for instance, is a notorious carrier of bacteria, especially under the food-storage conditions that prevailed in ancient times. Equally beneficial were the prohibitions against eating animals that were already dead, because these might be diseased. So, too, were the laws against consuming carnivorous (meat-eating) animals, which might carry toxins or poisons. On the other hand, virtually all fruits and vegetables, which are nutritious and less prone to carrying diseases than meat, are kosher (permitted) under Jewish law.

The diet of kosher foods helped make the Jewish community more close-knit and simultaneously separated Jews from their neighbors. Jews tended to congregate in groups large enough to ensure a reliable source of kosher food, and the Jewish diet led to the creation of an entire industry of butchers, bakers, and other tradespeople producing kosher food. This tradition continues today, and most large grocery stores in the United States include a kosher food section. Many Gentiles (non-Jews) also buy kosher products—for instance, kosher hot-dogs, which are considered healthier than the nonkosher variety.

Life Science

ESSAYS

away as China. Some of the more exotic drugs, to judge from the names by which they were listed in the medical records, include hippopotamus fat, fried mice, pigs' brains, and tortoise gall (bile).

Because their civilization flourished at a time when people had only a crude understanding of health and disease, many of the Egyptians' ideas concerning medicine were based on little more than superstition—and some of those ideas were actually dangerous. For instance, they considered the breakdown of food in the intestines to be a form of internal decay that could become a source of disease. Today, of course, everyone knows that without the breakdown, or metabolism, of food, the human body would have no means of receiving sufficient energy and a person would eventually die.

Even worse, the Egyptians thought that placing a foul-smelling substance (for instance, alligator excrement) on a wound would scare away evil spirits. Instead, the bacteria and other microorganisms in the dung most likely caused infection in persons unfortunate enough to receive this form of treatment. Nonetheless, the doctors of Egypt gained some skill at treating wounds, and though scientists possessed no concept of bacteria until the 1800s, the Egyptians' embalming techniques effectively combated the decay brought on by microorganisms.

Egyptian doctors also showed an understanding of the importance of documentation—that is, a written record of a patient's condition and the measures taken to address it. The Egyptians wrote on the fibers of a plant called the papyrus, whose name is preserved in the modern English word "paper." Thanks to surviving medical papyri (plural of *papyrus*), modern historians have written accounts of observations, diagnoses, and treatments, including surgery, undertaken by Egyptian physicians. Many of these cases involved the closing of wounds with stitches and adhesive bandages.

As one of the only ancient societies in which people were not forbidden to touch a dead body, Egypt most likely influenced studies in anatomy by physicians in later civilizations. Another influential aspect of Egyptian work in the life sciences was their use of a great wonder drug, a substance mentioned often in their papyri: honey.

Honey helps prevent the growth of bacteria. A twenty-five-hundred-year-old sample of honey unearthed from an Egyptian tomb, for example, showed almost no signs of decay. Like natron, it acts as a drying agent, drawing water from cells and causing them to die; and its uses in preventing decay made it popular with embalmers. Likewise, physicians valued it because it provided a protective coating between a wound and a bandage.

The Greeks and the Romans adopted the use of honey as a medicine, and even today, scientists are investigating its effectiveness in fighting certain

Life Science

ESSAYS

Ancient Egyptian druggists making an herbal remedy. (© Bettmann/Corbis. Reproduced by permission of the Corbis Corporation.)

types of bacteria that are resistant to standard antibiotic drugs. The medicinal use of honey was not the only way in which the Egyptians were ahead of their time. Only in the 1800s did European doctors finally adopt the Egyptian practice of using adhesive tapes to close wounds. Today this is a common practice both in homes (Band-Aids, for instance) and in hospitals.

☐ "ALTERNATIVE" MEDICINE BEGINS IN THE EAST

During the 1960s and 1970s, people in the United States and other Western countries became interested in the culture of the East or the Orient, particularly India. Celebrities such as the Beatles sparked an interest in Indian music, clothing, and religion. In the years that followed, many in the West found spiritual meaning in the philosophies and faiths of east Asia such as Hinduism; Buddhism, which began in India and spread to China; and Taoism, a religion native to China.

Widespread interest in Eastern religion had faded by the 1980s, but the West remains intrigued by Eastern methods of healing and maintaining health. Among such methods are acupuncture, which originated in ancient

Words to Know

acupuncture: The insertion of thin needles into specific points of the body in order to relieve pain or treat illness.

alternative medicine: Medical practices that are not officially recognized by the mainstream medical community.

Ayurvedic: A term describing a form of medicine, closely tied with the Hindu religion, practiced from ancient times in India.

China; yoga, an Indian belief system based on the harmony of body and mind; and various other alternative-health treatments.

The term *alternative medicine* refers to those practices that are not officially recognized by mainstream medical organizations—for example, the American Medical Association (AMA). However, in ancient China and India, where many of these ideas had their beginnings, alternative medicine was in the mainstream.

Acupuncture in China

Acupuncture, the insertion of thin needles into specific points on the body in order to relieve pain or treat illness, developed in China more than four thousand years ago. It is based on the idea that the body contains an essential life energy known as *qi* (pronounced "chee"), which flows in channels, or meridians. *Qi* is made up of two varieties of energy, *yin* and *yang*. Associated with female and male characteristics, *yin* and *yang* are opposite and complementary, meaning that the balance between them must be maintained. In the body, the male or *yang* elements indicate the capacity for activity and transformation, while the female or *yin* characteristics are associated with circulation, nourishment, and growth.

The aim of the acupuncturist is to remove barriers to the flow of *qi* and restore the balance of *yin* and *yang*, goals achieved by the placement of needles at certain positions on the surface of the patient's body. Like modern doctors, acupuncturists of ancient times usually began by asking the patient questions and making specific physical observations. They then set about treating the patient's condition by inserting needles at acupuncture points along the meridians.

Life Science

ESSAYS

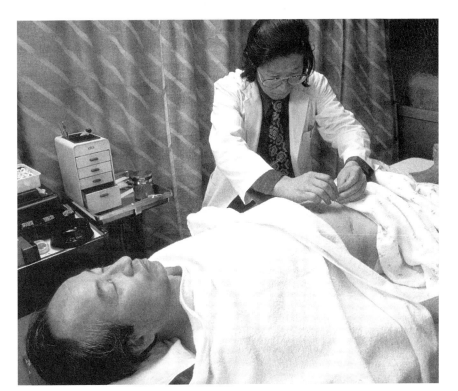

A doctor of traditional Chinese medicine performs acupuncture on a patient. Some Eastern forms of medicine, such as acupuncture, are gaining popularity in the West. (Reproduced by permission of Charles Gupton/The Stock Market.)

Acupuncture gains acceptance

Though it had originated long before, acupuncture only became widely accepted in China during the Warring States period (475–221 B.C.E.), a time of intense infighting throughout China.

Closely tied with this growing acceptance was the spreading influence of Confucianism and Taoism (DOW-izm), belief systems based on the teachings of Confucius (K'ung Cch'iu; 551–479 B.C.E.) and Lao-Tzu (c. 600 B.C.E.) respectively. Confucianism held that the body was sacred and that it should be presented whole—without any missing parts—to one's ancestors after death. It thus became very important to die with a "whole" body, and therefore the Chinese had a particularly strong fear of amputation, or the removal of a limb or other body part. Acupuncture offered a welcome means of treating internal disease while maintaining an intact body. Taoism, which placed an emphasis on balancing *yin* and *yang*, also fit well with acupuncture.

Yet another circumstance that influenced acceptance of this medical practice was the upheaval that followed the decline of the Han Dynasty. Established in 206 B.C.E., the Han Dynasty flourished until about 9 C.E., when the short-lived Hsin Dynasty overthrew the Han. Though the Han

Life Science

ESSAYS

The yin-yang symbol. Yin and yang represent two types of energy that are opposite and complementary, meaning that the balance between them must be maintained. (Reproduced by permission of Archive Photos, Inc.)

regained power in 23 C.E. and ruled for two more centuries, their influence was never again as great as it had been. The disorder in Chinese society continued after the overthrow of the last Han ruler in 220. Unlike most dynasties in Chinese history, the Han were not replaced by a stable government; instead, stability did not return to the country until the establishment of the Sui Dynasty in 581. This state of anarchy (lack of government) had its benefits, however, because it swept away many outmoded ideas and opened people's minds to new concepts such as acupuncture.

China and India; Buddhism and Hinduism

Though China and India are the two most populous nations in the world, and in premodern times they were the two most influential countries of east Asia, the Chinese and the Indians were unaware of one another's existence for many centuries. High mountains and wide deserts separate the two countries. Since the two cultures had almost no contact until about 500 C.E., they developed along quite separate lines.

Life Science
ESSAYS

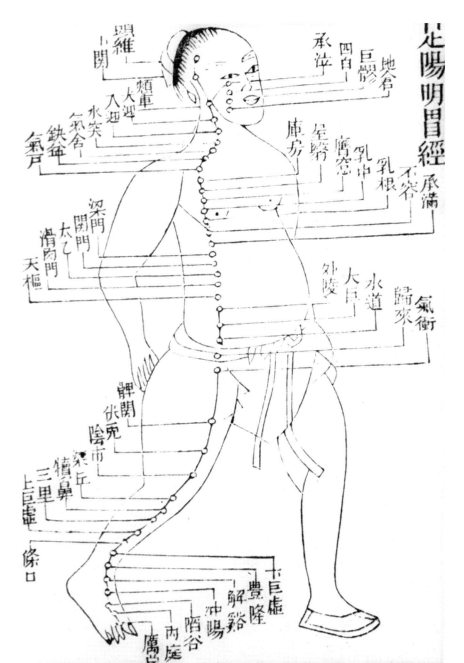

A chart showing acupuncture points. Acupuncturists insert needles at certain points on a patient's body in order to restore the balance of yin and yang. (Reproduced by permission of NLM.)

The establishment of Chinese-Indian contact was primarily due to the efforts of Chinese Buddhist monks, or holy men, who traveled to the birthplace of their religion—India. Buddhism is based on the principles taught by Gautama Siddhartha (GOW-tuh-muh si-DAR-tuh), or Buddha

Life Science

ESSAYS

(c. 563–c. 483 B.C.E.). Though Buddhism originated in India, it took hold much more strongly in China and other parts of east Asia.

India itself remained primarily under the influence of Hinduism, though parts of the population accepted Buddhism. Hinduism had been brought to the subcontinent by the Aryans, a people from the north who invaded India in about 1500 B.C.E. At that time, the religion was known as Vedism, the word *veda* (VAY-duh) meaning "sacred love." Associated with this religion was a set of writings called the *Ayurveda* (Science of life) that addressed various subjects—including medicine.

Principles of Ayurvedic medicine

Probably composed between 1400 and 1200 B.C.E., the ancient text the *Ayurveda* became the basis of Ayurvedic medicine. A medical system closely tied with Hinduism, Ayurvedic medicine emphasizes both physical and spiritual components of health. Also important were three other medical texts by the physicians, authors, and founders of the Ayurvedic system of healing, Charaka, Susruta, and Vagbhata. Their writings are known respectively as the *Charaka Samhita, Susruta Samhita,* and *Astangahrdaya Samhita,* though they are often designated in abbreviated form (the *Charaka, Susruta,* and the *Astangahrdaya*). The first two were probably composed between 1000 and 500 B.C.E., and the Astangahrdaya much later—most likely in the 600s C.E.

The *Susruta* focuses on surgery. Most operations performed by Indian surgeons involved the treatment of wounds, simply because the causes of wounds were usually quite visible—for example, an arrow stuck in a thigh. The book describes more than one hundred different surgical instruments, most made of iron, designed to look like the heads of specific animals. For example, there were forceps (surgical tongs) shaped like the mouths of crocodiles or herons. The basic designs of many of these ancient Indian surgical tools were very much like those still in use today.

The *Charaka* is primarily a classification of medicines. Ayurvedic doctors had at their disposal some one thousand drugs derived from plant sources, though a number of these are either unidentifiable or perhaps legendary. As for the work attributed to Vagbhata, it was based on the *Susruta* and *Charaka,* and is sometimes called *The Collection of the Essence of the Eight Limbs of Ayurveda.* These eight branches of medicine, also addressed in the *Charaka,* include areas such as diagnosing illnesses and prescribing drugs.

Similarities with Chinese and Greek medicine

Ayurvedic medicine was concerned primarily with maintaining a person's health rather than the treatment of disease. Taking an approach not unlike that of Chinese acupuncturists, Indian practitioners believed that the body incorporated a complex system of "vital points"—places where major veins, arteries, ligaments, joints, and muscles unite—called *marmas*. The locations of *marmas* and acupuncture points on the body are quite similar. Unlike the Chinese, however, Indian physicians were more likely to prescribe amputation, particularly if an injury would otherwise be fatal due to the *marma* involved.

As was the case with the Greeks, Indian physicians of ancient times subscribed to the idea of humors (see essay "Greek Physicians Transform the Life Sciences" in this chapter). Instead of four humors, however, the Indians believed in three: *vata, pitta,* and *kapha*—which are usually translated as wind, bile, and phlegm. Also like the Greeks, Ayurvedic physicians believed that the body was composed of only a few elements: earth, water, fire, and wind (all components of early Greek ideas about science), and a fifth element, empty space.

Humors and elements (at least, in the sense described here) are now regarded as wholly unscientific, as is much else about Ayurvedic medicine. Like many early approaches to the health sciences, it associated illness with spiritual conditions such as demon possession. Furthermore, as was the case with European doctors almost to the beginning of modern times, ancient Indian physicians believed that it was sometimes necessary to remove blood from a patient in order to restore balance in the body. Despite these limitations, however, Ayurvedic medicine achieved results in areas—such as smallpox inoculation—that would not be equaled in the West until the late modern era.

Medicine and social problems in India
Perhaps Ayurvedic medicine's greatest contribution was its social impact. For instance, the Indian method of smallpox inoculation represented one of the first attempts to deal with a large-scale epidemic. Furthermore, ancient Indian doctors made strides in plastic surgery, which addresses social and psychological conditions (that is, a person's feelings about how he or she looks) to an even greater extent than it does purely health-related problems.

The Ayurvedic method of reconstructing noses was particularly significant. Indian soldiers' helmets did not protect their noses from injury by enemies' swords; furthermore, people who had been convicted of adultery (having sexual intercourse with someone other than one's spouse) were punished by removal of the nose. Ayurvedic doctors addressed this problem by cutting a roughly nose-shaped piece of skin from the patient's forehead or cheek. This flap of skin would be left attached along one edge so that blood would continue to flow to it; otherwise, the cut skin would die.

Life Science

ESSAYS

Achievements of Ancient Indian Medicine

A girl receiving a smallpox vaccination. Among the many early achievements of ancient Indian medicine was inoculation for smallpox. (© Corbis-Bettmann. Reproduced by permission of the Corbis Corporation.)

A number of treatments used by Ayurvedic doctors of ancient India showed that these physicians were ahead of their time. For example, for snakebites they applied a tourniquet, thus slowing the spread of poison by cutting off circulation. These ancient doctors' knowledge of this technique is all the more impressive since they had no concept of how blood circulates throughout the body. Even more amazing, however, was their method of treating smallpox, which remained one of the world's most deadly diseases until its elimination in 1978.

Indian physicians apparently took pus or scabs from the sores of a mildly infected patient, or from a cow infected with a related disease called cowpox, and rubbed the material into a small cut made in the skin of a healthy person. Clearly this idea, known as inoculation, goes against common sense—but it works, as modern efforts have proven. The Indians' method was a risky one, and there was always a chance that the patient would

Next, the surgeon rotated the flap of skin and stitched it into position to form a new nose. British surgeons later adopted this method from Indian physicians during the eighteenth century, and today's procedures for nose reconstruction are still based on it.

Ayurvedic medicine also helped physicians overcome several social taboos, such as the restriction—common in most ancient societies—

become deathly ill. The problem with the Indian method of inoculation was that Indian doctors sometimes used smallpox, rather than cowpox, to inoculate patients. Smallpox is much more deadly, and because they did not always regulate the amounts administered to patients, many died from the inoculation process. In England many centuries later, however, Edward Jenner (1749–1823) realized that cowpox was every bit as effective as smallpox for inoculating patients against the latter disease, and administered the vaccine in highly regulated doses. As a result, few people suffered from the inoculation process, and the threat of smallpox began gradually to recede.

When it came to surgery, ancient Indian physicians conducted some of the most advanced work in the world prior to modern times. If a pregnant woman experienced complications in childbirth and died on the operating table, the doctor would cut open the mother's abdomen and uterus and remove the baby—an early form of cesarean section (see essay "The Medical Discovery of Women" in this chapter).

Indian physicians also developed a technique for the removal of bladder stones, crystals formed from certain substances in urine. These can be very painful and pose a serious threat to a patient's health because they often grow large enough to block the passage of urine. The Ayurvedic treatment involved locating the stone by touch through the outside of the bladder wall, then making a small incision or cut through which the stone could be removed. Greek and Roman physicians later adopted this technique, which was practiced in Europe and Asia until the nineteenth century.

against dissecting human bodies. In order to get around the religious prohibitions against using a knife on the dead, the *Susruta Samhita* used a method for covering a dead body with various grasses, placing it in a cage of mesh, and letting it soak for a week in a pond. Then the upper layers of skin could be removed by gently rubbing the body with soft brushes, and the doctor could observe the inner parts of the cadaver, or dead body.

Religion, medicine, and the caste system

In addition, Ayurvedic medicine addressed a social taboo unique to India: the caste system. Associated with the Hindu belief in reincarnation, the caste system divided people into rigidly defined social classes. A person could not simply work his or her way out of one caste into a higher one. By definition, people in a higher caste were believed to have evolved to a higher spiritual state through repeated birth. Thus the group known as the Brahmans, who occupied the highest levels in terms of wealth, power, and prestige, were also considered the most spiritually advanced. At the bottom rung were the Harijan, or Untouchables, who, as their name implies, could not be allowed contact with persons of higher caste.

This, of course, presented problems where proper medical treatment was concerned—not least because it was often people at the lowest social levels who most needed a doctors' care. Under Ayurvedic medicine, many of the old caste restrictions remained in place; however, physicians were free to accept students from all three of the upper castes. In areas controlled by Buddhists, however, no caste restrictions existed, and during the period of the Gupta Empire (c. 320–c. 540), which united much of northern India into a cohesive political and religious entity, Buddhist monks in India established some of the world's first hospitals—facilities that treated the poor for free.

Chinese, Indian, and alternative medicine today

Though modern Indian scientists and doctors have adopted Western science and medicine, millions of people in India still use Ayurvedic remedies. The Indian Medical Council, which was established in 1971, recognizes various forms of traditional medical practice and sponsors attempts to combine Indian and Western forms of medicine. In China, acupuncture remains an active form of medical treatment, and Chinese communities in other countries have spread this form of "alternative" medicine to the rest of the world.

The mainstream medical community in the West remains skeptical regarding acupuncture and other alternative medical treatments, some of which (for example, chiropractic, a system of therapy that uses manipulation and body adjustments to fight illness and disease) are Western rather than Eastern in origin. This skepticism comes primarily from the fact that many alternative forms of healing address spiritual matters that are simply beyond the scope of science. It is therefore difficult to apply a regular system of observation and experimentation—which is essential to scientific study—where alternative medicine is concerned. Nonetheless, millions of people in both East and West continue to believe in, and in many cases receive health benefits from, forms of medicine that are far from the Western mainstream.

Words to Know

caduceus: A staff with two entwined snakes and two wings at the top, which is the internationally recognized symbol of medicine.

dissection: Cutting open dead bodies in order to study their internal characteristics.

humors: Four fluids (phlegm, blood, yellow bile, and black bile) that, according to Hippocrates and his followers, made up the human body. Imbalances in these humors were supposedly responsible for all illnesses.

physiology: The scientific study of the functions, activities, and processes of living things.

prognosis: A prediction regarding the course and outcome of a disease based upon previous observation of similar cases.

GREEK PHYSICIANS TRANSFORM THE LIFE SCIENCES

Though other civilizations, most notably those of Egypt and India, made important contributions to the study of the life sciences, these are overshadowed by the many achievements that took place in Greece and in Greek-influenced lands between about 500 B.C.E. and 200 C.E. Particularly significant was the work of two physicians, **Hippocrates** (c. 460–c. 377 B.C.E.; see biography in this chapter), whose work gained prominence at the start of this era, and **Galen** (129–c. 199; see biography in this chapter), whose career coincided with its decline.

Historians know little about Greek medicine prior to the "golden age" in the 400s B.C.E., when Greek civilization was at its peak, though it appears that the earliest Greeks used folk healing and herbal drug therapies. Healers often summoned help from gods, most notably Asclepias (as-kluh-PY-us), son of Apollo and god of medicine. Asclepias was usually portrayed holding a staff with two entwined snakes and two wings at the top—a symbol known as the *caduceus* (kuh-DOO-see-us), which today represents medicine throughout the world.

Life Science

ESSAYS

Hippocrates and his followers

One of the most prominent figures of the Greek golden age was Hippocrates (hi-PAHK-ruh-teez), known as the "father of medicine." Prior to Hippocrates, medicine had always been linked with religion, but Hippocrates defined disease scientifically, according to physical rather than spiritual causes. In addition, he was the first scientist to accurately describe the symptoms of several diseases, including pneumonia and epilepsy in children.

A nineteenth-century caduceus. This symbol of the medical profession dates to the fifth century B.C.E. (Corbis-Bettmann. Reproduced by permission of the Corbis Corporation.)

Teaching that the body has a powerful ability to heal itself naturally, Hippocrates often prescribed treatments of rest, fresh air, and bathing. Many of his therapies involved changes in diet; on the other hand, he was reluctant to recommend surgery. Hippocrates developed the concept of *prognosis*, a prediction regarding the course and outcome of a disease based upon previous observation of similar cases. By seemingly "foretelling the future," he and his followers gained the trust of their patients and thus replaced traditional healers as figures of influence in Greek society.

The Hippocratic corpus and oath

In ancient Greece, it was typical for a great thinker to establish a school—not necessarily a physical building, but rather a group of students who would continue and in many cases expand upon their master's work. In the case of Hippocrates, it is possible that some of the achievements attributed to him were actually the work of his students.

Such is the case with the Hippocratic corpus, a set of writings traditionally credited to Hippocrates. Regardless of the author's identity, the Hippocratic corpus was the first known medical library. The Hippocratic corpus stressed the idea that health was a matter of equilibrium or balance, and that illness resulted from an upset of that equilibrium.

Much more famous than the Hippocratic corpus is the Hippocratic oath, derived from the corpus, which established a code of ethics (moral behavior) for doctors. In its first half, the oath addresses the physician's responsibilities to his or her teachers and students, while the second half outlines rules for personal and professional conduct—in particular, the physicians' treatment of patients.

Life Science

ESSAYS

The Hippocratic Oath established a code of ethics for doctors. (© Corbis. Reproduced by permission of the Corbis Corporation.)

The four humors

Among the most influential ideas in Greek medicine, embraced by Hippocrates and his followers as well as many others, was the doctrine of the four humors. This theory had its roots in the work of the Greek philosopher Empedocles (c. 490–430 B.C.E.), who described nature as being made up of four elements: earth, air, fire, and water. Hippocratic medicine reflected these elements in the four fluids, or *humors,* that supposedly made up the human body: phlegm (FLEM), blood, yellow bile, and black bile. (Bile is a fluid secreted by the liver.)

Life Science

ESSAYS

According to humoral theory, the warm wetness of earth was related to blood; the cold dryness of air to black bile; the hot dryness of fire to yellow bile; and the cold wetness of water to phlegm. All diseases were said to be the result of imbalances in these four humors. When a person caught a cold, for instance, this meant that he or she had an excess of phlegm.

Challenges to humoral theory

In the years following Hippocrates's death, many scientists and physicians accepted, and contributed to, humoral theory. One of the few who rejected it was, interestingly, a man named after the god of medicine: Asclepiades of Bithynia (as-kluh-PY-uh-deez; c. 100 B.C.E.). Asclepiades subscribed to the idea that all of nature is composed of atoms—itself another significant contribution of ancient Greek thought. In accordance with this, he maintained that illness resulted from interruption in the motions of atoms.

Though still incorrect, Asclepiades's ideas were far more accurate than the doctrine of the four humors, which had become firmly established by the beginning of the Middle Ages in about 500 C.E. One of the unfortunate results of humoral theory was the notorious practice of bloodletting. According to the theory, certain illnesses resulted from an excess of blood; therefore doctors would often attach blood-sucking leeches to a patient's body. Today, of course, physicians realize that in most cases it is not beneficial for a sick person to lose blood.

Other humoral treatments, such as purging (induced vomiting), starvation, or the forced consumption of liquid, often contributed to patients' ill health. Less harmful was the association of humoral theory with a crude form of psychology, from which the modern meaning of "humor" is taken (see sidebar).

Galen and the birth of physiology

Galen (GAY-lun) is remembered as the "father of physiology," the scientific study of the functions, activities, and processes of living things. Like Hippocrates, he was Greek, but by the time of Galen's birth—some six hundred years after Hippocrates—most of Europe had fallen under the control of the Roman Empire.

Galen himself served as court physician to the Roman emperor Marcus Aurelius, whose death in 180 C.E. marked the beginning of Rome's long decline. Three centuries later, in 476, the Roman Empire's control over western Europe would come to an end, but by then the progress of science in the West had long since ground to a halt. Galen, who wrote numerous medical texts, was thus the last great doctor in western Europe for at least a thousand years.

Humor, the Humors, and Personality Types

The idea of the four humors, or bodily fluids, originated in ancient Greece and continued to dominate Western medicine until almost 1500. Humoral theory had a number of negative effects, such as encouraging the practice of bloodletting. The humors also became associated with four basic personality types. The term "humor" itself described the amusing behavior exhibited by someone who supposedly had too much of a given humor in his or her body.

Even today, people use the terminology of the four humors to describe basic personality types. A choleric (kuh-LAYR-ik), once believed to be an individual with an excess of yellow bile, is someone who likes to be in control—a person who has his or her own way of doing things and expects others to get in line. The opposite of a choleric is a phlegmatic, an easygoing and seemingly unflappable person. The sanguine and melancholy personalities are also opposites. The sanguine, once associated with a surplus of blood, describes a person who loves to have fun but is not particularly well organized. A melancholy personality (formerly equated with an excess of black bile) is sad and brooding, often artistic.

Of the characters in the *Winnie-the-Pooh* stories created by British author A. A. Milne, for instance, Pooh himself would probably be classified as a phlegmatic. Tigger would almost certainly be sanguine, while Kanga and Rabbit appear to be choleric. Piglet and (of course) Eeyore are melancholy. Few people or characters can be described as pure types, however. Pooh has secondary sanguine characteristics, and Eeyore, for all his grumbling, often displays the peaceful acceptance of a phlegmatic.

The comparison to *Winnie-the-Pooh* has been chosen deliberately, not only because those are stories of almost universal appeal, but also to emphasize the fact that the idea of the four personality types is about as scientific as the Pooh stories themselves. Nonetheless, analysis of the four humors provides a fun and sometimes revealing way of looking at people, characters, and oneself.

CONCORDIA COLLEGE LIBRARY
BRONXVILLE, NY 10708

Life Science

ESSAYS

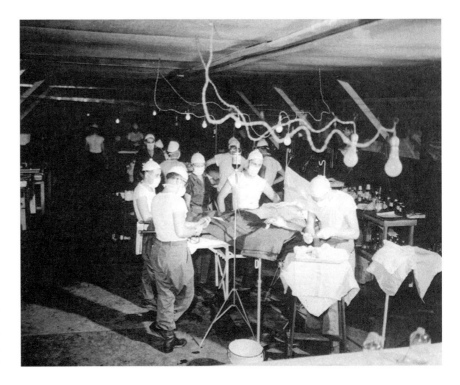

A Mobile Army Surgical Hospital (MASH) near the front during the Korean War. The Romans first developed military surgical units to treat wounds on the battlefield. (© National Archives and Records Administration. Reproduced by permission of Double Delta Industries.)

Truth and superstition

Dissection—cutting open dead bodies in order to study their internal characteristics—was key to much of Galen's work. At that time, religious law forbade the dissection of human bodies, and therefore Galen could cut open only the carcasses of animals. Because he was studying animal anatomy rather than human anatomy, he drew some questionable conclusions.

Galen's investigations led to his discovery that the arteries carry blood, not air, as earlier scientists and physicians had believed. He correctly noted that the liver removes waste products from the bloodstream but went on to claim that the liver transforms food into blood, which he believed then moved to various organs. By bringing together blood, air, and food, he paved the way for the highly significant studies in blood circulation conducted by English physician **William Harvey** (1578–1657; see biography in this chapter) over a thousand years later. Harvey, however, showed that Galen was incorrect in assuming that the liver converts food.

Galen also subscribed to "pneumatic theory," which combined a trace of science—the idea that air (*pneuma*) is vital to the functioning of the human body—with unscientific notions, including the idea of psychic spirits living in the brain. Yet Galen had the ability to draw truth from superstition. His pneumatic studies, for instance, led to the discovery of the fact that an individual's pulse is a key indicator of that person's overall health. He also correctly noted that muscles always work in pairs, and he advanced the cause of science with his emphasis on experimentation as a critical component of scientific work. He wrote hundreds of books on medicine, which became the basis for medical education in the Muslim world and later in western Europe.

Galen's ideas remained dominant until Belgian anatomist **Andreas Vesalius** (1514–1564; see biography in this chapter) successfully showed that while, while many of Galen's teachings on anatomy were correct with regard to certain animals, but they did not describe human physiology.

The establishment of hospitals

Many of the mistakes made by the great thinkers of ancient times would not seem so glaring today if their ideas had not held such a powerful influence for such a long time. It is not their fault, however, that learning came to a virtual halt with the decline of the Roman Empire, and that western Europe produced few scientists of note for a millennium.

The most significant development in medicine after the fall of the Roman Empire, was the establishment of hospitals, which had their roots in the military medical facilities of Rome. Indeed, the Western concept of public health originated with the Romans, who greatly surpassed the Greeks in precautions relating to sanitation, sewage, and drainage to help reduce the threat of epidemics.

Because the Romans spent many centuries fighting wars of conquest, it became necessary to develop a means of treating wounds on the battlefield. Thus was born the concept of the military surgical unit, facilities—sometimes temporary and mobile, sometimes permanent—that dotted the Roman world. Galen himself was a doctor at one such hospital.

Christian hospitals

As Rome declined and Christianity replaced the old pagan Roman faith, the establishment of Christian hospitals gave Christianity a great appeal among the poor and the sick. As early as 250 C.E., the bishop of Rome (the pope) had established houses of refuge for people who suffered from the skin ailment known as leprosy.

The first true hospital was probably the one established near the Greek town of Caesarea in 369 C.E., and perhaps the most significant founder of hospitals in ancient times was the Roman noblewoman Fabiola (died c. 399). A woman of wealth and influence, she is said to have physically carried the poor and the sick off the streets, washing and caring for

Life Science

ESSAYS

Life Science

ESSAYS

them herself. In the centuries that followed, important hospitals were established in areas from England to Asia Minor (modern Turkey).

☐ THE GREAT DOCTORS OF MEDIEVAL ISLAM

The decline of the Roman Empire from about 200 to 500 C.E. left a power vacuum in the Mediterranean region that would not be filled until the mid-600s. The only two important forces in the area were the Byzantine Empire in Greece and the Persian Empire in what is now Iran, but neither equaled Rome's former power. Yet in the seventh century, tribes from Arabia would emerge to build an empire as great as that of the Romans.

Driven by the Muslim faith, which was based on the teachings of the prophet Muhammad (c. 570–632), the Arabs became the leaders of a revolution that would sweep the world. Within a century of Muhammad's death, Muslim armies had conquered a vast empire that stretched from Spain to what is now Afghanistan. In the process, they removed the Persian rulers, took Byzantine holdings in north Africa and the Middle East, and very nearly overran western Europe.

But the Arabs were more than simply military conquerors. Most of the scientific advances from about 800 to 1200 occurred at Muslim centers of learning in Cairo, Damascus, and Baghdad (now the capitals of Egypt, Syria, and Iraq respectively), as well as further east in Persia (Iran) and central Asia. Islamic scholars such as **Rhazes** (c. 865–c. 923; or ar-Razi; see biography in this chapter); **Avicenna** (980–1037; or Ibn Sina; see biography in this chapter); and **Averroës** (1126–1198; or Ibn Rushd; see biography in this chapter) were among the most advanced thinkers in the world, and their writings furthered scientific learning throughout the Middle Ages (roughly 500 to 1500).

Note that each man is referred to by two different names: first, the name by which he came to be known in Europe; and second, by an abbreviated form of his Arabic name. The Arabic names could be extremely long, whereas a name like Ibn Rushd is rather like referring to someone by his last name only. The term *Ibn* means "son of," similar to European names that end in -son, such as Johnson or Anderson. The prefixes *ar-* and *an-*, along with the more commonly used *al-*, are simply terms of respect, rather like "mister."

Peoples and languages of the Muslim world
Generally speaking, the peoples who occupy north Africa and a region from the borders of modern Turkey to the tip of the Arabian Peninsula (with the exception of Israel), are Arabs. As is the case with most ethnic

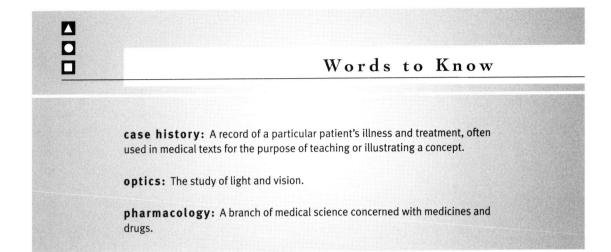

Words to Know

case history: A record of a particular patient's illness and treatment, often used in medical texts for the purpose of teaching or illustrating a concept.

optics: The study of light and vision.

pharmacology: A branch of medical science concerned with medicines and drugs.

groups, they are properly distinguished not by racial characteristics, but by their native language—in this case, Arabic. "Persian" specifically means someone from Iran, but in the medieval context the term referred to people in Iran, Afghanistan, and central Asia who spoke Farsi (Persian) and related languages.

During the Middle Ages, Arabic was a common language of learning in the Islamic world, just as Latin was in western Europe—and just as English is today. Farsi had the same role in the eastern lands controlled by Islam, but even there, educated people who were not Arabs still spoke Arabic. In any case, the controlling religion in the region, which provided a system that united diverse peoples, was Islam.

Thus the term "Muslim" is more all-encompassing than "Arab" to describe the great civilization of the Middle East in medieval times. Still, even that term does not convey the whole picture. In addition to contributions from Muslim Arab and Persian scientists, the intellectual life of the region benefited also from the work of Jewish scientists, as well as Middle Eastern Christians and members of other religious or ethnic groups.

But the ruling ethnic and linguistic group in the Muslim world were the Arabs, who spread out from Arabia (the Arabian Peninsula) to populate an area from Morocco to Iraq. At first the Arabs were indifferent to the learning and educational institutions of conquered peoples, but gradually they grew to appreciate these. By the middle of the 700s, they had begun to employ various translators to render scientific, medical, and philosophical works from Greek, Farsi, and Hindi (the language of northern India) into Arabic.

Life Science

ESSAYS

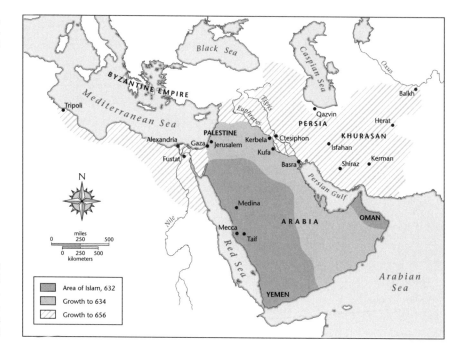

A map of Muslim holdings at the height of medieval Islamic power, c. 600. (Map by XNR Productions. Reproduced by permission of the Gale Group.)

Translations spur medical work

The Greek manuscripts, captured from former Byzantine territories such as Egypt, would prove to have the greatest influence, and medicine became the first of the Greek sciences to receive attention. In this effort, scholars had the blessing of the caliphs, the spiritual and political leaders of the Islamic world, who ruled from Baghdad. There the caliphs had established a translation bureau called the House of Wisdom, and they sent scholars to Greece and former Greek lands to purchase ancient scientific works. At the same time, they gathered the knowledge of Persia, India, and China, and thus ushered in the first era of Islamic medicine—the period of translation and compilation.

During this time, Islamic physicians simply preserved ancient knowledge by collecting, then translating, the classic writings of Greek thinkers such as **Hippocrates** (c. 460–c. 377 B.C.E.; see biography in this chapter), **Aristotle** (384–322 B.C.E.; see biography in this chapter), and **Galen** (129–c. 199; see biography in this chapter). However, they soon went beyond the ancients, adding information from other cultures (primarily India and China) along with their own comments and interpretations.

From Greek medicine to Islamic medicine

Islamic medicine largely accepted the humoral theory of Hippocrates and Galen, who maintained that health is governed by four bodily fluids, or humors: blood, phlegm, black bile, and yellow bile. When the humors were in perfect balance, the body was healthy, but when the humors were out of balance, the person became ill. In accordance with these ideas, physicians viewed it as their task to preserve health if possible, and to heal if necessary.

Muslim thinkers thus began by accepting the wisdom of the ancients without question, but eventually they challenged the ideas of the Greek masters and in the process laid the foundations for modern medicine. Particularly important was the work of Muslim physicians in pharmacology, the study of drugs and medicines, and optics, which is concerned with light and vision. The discovery and translation of ancient classics thus paved the way for a blossoming of the life sciences, reflected throughout the Muslim world by the establishment of hospitals, apothecary shops (pharmacies), and medical schools.

Great Islamic medical writers
During a period of several centuries, the Islamic world produced a number of brilliant scientists and physicians. Of these, several are particularly notable, in large part because they left behind extensive writings—not just translations, but original works.

Among these were the Persian physician Rhazes, the Persian physician and philosopher Avicenna, the Spanish Arab physician and philosopher Averroës, and the Egyptian physician Ibn an-Nafis (died 1288).

Rhazes
Rhazes (c. 865–c. 923) is best remembered for his encyclopedia of medicine and for *A Treatise on Smallpox and Measles,* a pioneering work in which he became the first scientist to differentiate between the two illnesses. The encyclopedia, *al-Hawi,* appeared in Latin in 1279 as *Liber continens* (Comprehensive book), and remained the principal textbook on medical therapy in the West for three centuries.

Rhazes wrote an additional two hundred books and papers and claimed to have written twenty thousand pages in just one year, yet he also found time to serve as first physician-in-chief for the prestigious Audidi hospital in Baghdad. Through his supervision of the hospital, as well as in his private practice, Rhazes compiled many intriguing case histories (a record of illness and treatment used for the purpose of illustrating or teaching a concept) that reveal a great deal about the conditions and techniques common at the time.

Avicenna
The English title of Avicenna's greatest work is the *Canon of Medicine,* which dominated medical teaching in Europe and appeared in countless

Life Science
ESSAYS

Life Science

ESSAYS

A painting from the Arabic manuscript De materia Media *showing an apothecary stirring a large cauldron while a patient looks on. Islamic medical writers left behind extensive writings on their medical practices. (Reproduced by permission of the Granger Collection, New York.)*

languages. The medical text remained in use for over eight hundred years—well into the nineteenth century in some areas. At more than one million words (nearly three thousand published pages), it represents a comprehensive collection of medical knowledge from the Greek, Indian, Persian, and Arab civilizations.

In this vast work, Avicenna used Galen and the humoral theory as starting points, but went much further. Stressing the importance of prevention, he discussed a number of diseases and a variety of factors that cause them. The anatomical observations in the *Canon* were exceedingly detailed: for instance, the book discusses at great length the dilation and contraction (widening and narrowing) of the eye's pupils, as well as the six motor muscles of the eye and the function of the tear ducts.

In addition to describing 740 medicinal plants and the drugs made from them, Avicenna laid out a series of basic rules for clinical drug trials, many of which continue to serve as the basis for testing new drugs today.

For instance, he suggested what would now be called a controlled experiment, by testing the drug on at least two different types of diseases. He also called for what is known today as targeted treatment, arguing that the drug must be applied for a "simple" ailment, not one with several complications.

Condemning mere guesswork in anatomy, Avicenna encouraged observation and a close study of the human body. Several of his findings, such as his suggestion that nerves cause pain and enable muscle movement, were ahead of their time. Clearly written and organized, the *Canon* became perhaps the most influential medical work of all time, and Avicenna—who wrote some twenty other medical books—became a more popular medical writer than even Galen.

Averroës and Ibn an-Nafis
Though Avicenna and Averroës (uh-VEER-oh-weez) are often spoken of together, and both were physician-philosophers, the first is remembered primarily for his medical work and the second for his writing on philosophy. Just as Avicenna's writing became more popular than Galen's did, scholars preferred Averroës's commentaries on Aristotle—for which he earned the nickname "the Commentator"—to those of Aristotle himself. Averroës wrote more than twenty works on medicine, including *Kulliyat* (General medicine), a textbook modeled on Avicenna's *Canon*.

Ibn an-Nafis, who lived almost four centuries after Rhazes, represented the latter part of the Islamic cultural boom. He was the first to describe pulmonary circulation, or the flow of blood between the heart and the lungs, but contemporaries ignored his discovery. His work gained attention only in the twentieth century. Galen had recognized that blood passes between the right and the left side of the heart, but he claimed that it did so by seeping through invisible pores in the septum, the wall that separates the two chambers. Ibn an-Nafis challenged this, indicating that in fact the blood flows from the right ventricle or chamber to the lungs, and from there to the left ventricle—findings later established by English physician **William Harvey** (1578–1657; see biography in this chapter).

Evaluating Muslim achievements
The writings of Islamic physicians and philosophers, often presented merely as commentaries on the work of Galen, would eventually be translated from Arabic into Latin and served as primary textbooks in European universities for hundreds of years. As such they helped greatly advance the life sciences in the West and were a major force behind Europe's emergence from the superstition and quackery that plagued the life sciences in medieval times.

Life Science

ESSAYS

Life Science

ESSAYS

But for many European scholars, medieval Arabic writers were significant only for their role in preserving Greek philosophy. Therefore the Arabic manuscripts that received the most attention were the ones that most closely followed the Greek originals. On the other hand, if a writer dared to question Galen, European translators often dismissed such as works as corruptions of ancient Greek wisdom. As a result, European scholars managed to confirm their own prejudice that Muslim scholarship lacked originality.

Until about 1950, Western historians tended to view medieval "Arabian" science, as they called it, purely as a postscript to ancient Greece. Since then, however, scholars have sought a broader view of what is now called "Islamic medicine." As part of this attempt, scholars have examined many Arabic manuscripts that were previously unknown, ignored, or unpublished.

Some historians believe that hundreds, or even thousands, of manuscripts written between about 1000 and 1500 are still gathering dust in various libraries throughout the world. Further studies of such manuscripts may clarify certain questions and perhaps lead to new discoveries regarding the achievements of Muslim scholars during the Middle Ages.

☐ FIFTEEN HUNDRED YEARS OF SUPERSTITION

During the Middle Ages (roughly 500 to 1500) and through the Renaissance (the period of artistic and intellectual rebirth, lasting roughly from 1350 to 1600, that marked the end of the Middle Ages), physicians who were educated and trained in the scientific method—with a reliance on observation and experimentation—were few in number. Most of the population relied on a combination of alternative healers—including empirics (those outside the medical mainstream), barber-surgeons, and apothecaries—for their medical care. These healers learned their trade mostly through apprenticeships and used folklore, herbs, and guesswork to cure the sick.

Disease was often attributed to supernatural causes, from evil spirits to punishment by God. Superstition and religious fervor dominated the philosophy of caring for the sick. In Europe the population embraced empirical healers along with the prevailing mysticism in desperate attempts to escape the great plagues of the late Middle Ages.

By 1661 as London was in the grip of another plague, new explanations for diseases were sought. Interest in science was rekindled and classical texts were rediscovered. As Renaissance thinking spread from Italy

Hospitals of the East

The word *hospital* comes from a Latin word meaning "guest." Indeed, the Romans established the first hospitals as medical facilities for their armies, but by the Middle Ages the concept had spread to the Muslim world as well. As in the Christian countries of western Europe and the Byzantine Empire (Greece and eastern Europe), the idea of a hospital was linked with religious notions of charity; like the Bible, the Koran, Islam's holy book, calls on believers to care for the sick and the poor.

The first important medical facility of the Middle East was in the Persian town of Jundishapur, whose hospital was associated with a university where scholars translated ancient Greek works into Arabic. In the centuries that followed, the Muslims built dozens of hospitals throughout their territories, including a particularly impressive one, the Audidi, in the capital city of Baghdad.

The Audidi's first administrator, **Rhazes** (ar-Razi; c. 865–c. 923; see biography in this chapter), allegedly selected the site by hanging pieces of meat at various places around the city: the place where the meat took the longest to rot, supposedly, had the best air quality. Another important Muslim facility was the Mansuri in Cairo, Egypt, built in 1283. It had male and female nurses; special wards for women and for patients with fevers, eye diseases, or mental illnesses; as well as its own pharmacy, library, and lecture hall.

The best hospital in the Byzantine Empire was the Pantocrator ("Christ the Ruler"), in the capital city of Constantinople (now Istanbul, Turkey). Completed in 1136, it was a huge complex of buildings divided into wards, much like a modern institution. By comparison, most hospitals in western Europe during the Middle Ages were little more than quarantine facilities, called pesthouses, to keep sick people away from the rest of the population.

Yet even the more sophisticated hospitals of the era bore little resemblance to modern facilities. For one thing, providing medical care was not their sole or primary purpose; instead, these medieval hospitals served as centers of religion, healing, faith, philosophy, *and* scientific medicine.

Words to Know

apothecary: One who prepared and sold medicines in medieval times.

astrology: An unscientific belief system based on the idea that the positions of the planets and the stars have an effect on human destiny.

herbalist: Someone who deals in medicinal herbs. Herbalists played a major role in medieval medicine and continue to do so in alternative medicine today.

life expectancy: A calculation, based on statistical data, of the life span for the average person.

quack: Someone who falsely claims to possess medical skill and the ability to provide proper treatment.

throughout Europe bringing the rebirth of the scientific method based on observation and experimentation, the early foundations for modern medicine were laid. Slowly, empirics and folklore medicine declined, and the science of medicine gained prominence.

Among the conditions typical of even the healthiest people of the Middle Ages were skin diseases, rotting or missing teeth, and crooked bones (the result of poorly mended fractures, bad nutrition, or other ailments). Bathing was rare and used mostly as a form of medical therapy, while dental hygiene was virtually unknown.

Nor was the problem merely one of bad teeth and body odor: the life expectancy for the average person in the Middle Ages was only about thirty years. In the United States at the end of the twentieth century, in contrast, life expectancy at birth was nearly seventy-three years for the average male, and nearly eighty years for the average female. In medieval Europe, since many women died in childbirth, life expectancy was lower for women than for men. Life expectancy typically increases with age, as a person gets past life-threatening circumstances such as childhood diseases. Most families in the Middle Ages lost some members to diseases such as scarlet fever.

Life Science

ESSAYS

An illustration from a Renaissance period book by Johann Geller von Kaisersberg. Although the Renaissance was the beginning of modern history, a lot of medieval ideas and fears, such as those about werewolves, still circulated. (Reproduced by permission of the Mary Evans Picture Library.)

Putting the period into perspective

When the Roman Empire began to decline after 200 C.E., political instability and fears of invasion created an environment in which it was impossible to foster an active scientific community. Instead of moving forward, as it had done for centuries, science in Europe began to move backward.

After the empire permanently split into eastern and western halves in 395, eastern Europe remained fairly secure under what came to be known as the Byzantine Empire, ruled from Greece. Even there, however, doctors and scientists were mostly content to rest on the ideas of ancient physicians such as **Hippocrates** (c. 460–c. 377 B.C.E.; see biography in this chapter) and **Galen** (129–c. 199; see biography in this chapter). The Byzantines shut themselves off from western Europe, which continued to decline in the years leading up to the fall of the Western Roman Empire in 476.

That event ushered in a period of confusion and ignorance sometimes called the Dark Ages, which lasted until the beginning of the Crusades in 1095, a series of religious wars in which Europeans attempted to gain control over the Muslims who ruled the Middle East. The Crusades introduced Europeans to the sophisticated civilization of the Muslim world, and this in turn spurred the gradual reemergence of scientific learning in the West and paved the way for the period of rebirth known as the Renaissance.

Often cited as the beginning of modern history, the Renaissance occurred from about 1350 to 1600, overlapping the Middle Ages. Indeed, many of the primitive practices that people associate with medieval medicine—for instance, barbers performing surgery—were actually more a part of the Renaissance than of the Middle Ages.

Life Science

ESSAYS

Explanations of disease

Even the most qualified physicians, or "physics" as they were called, of the Middle Ages combined a great deal of superstition and mysticism with a smattering of learning based on the teachings of Hippocrates and Galen. Though those two were the greatest doctors of the ancient world, they were wrong about a number of things, and medieval physicians tended to adopt the worst ideas of Hippocrates and Galen.

Among these was the doctrine of the four humors, the belief that a balance of bodily fluids governs human health: blood, phlegm, yellow bile, and black bile. If a person had too much of any one humor—including blood—that excess fluid would have to be drained, for instance by placing blood-sucking leeches on a patient's body. In seeking the causes and cures for diseases, doctors also turned to *astrology,* an unscientific belief system based on the idea that the positions of the planets and the stars have an effect on human destiny.

Physicians of the Middle Ages did not seek to understand diseases in themselves but assumed that each person got sick in an entirely different way, and for different reasons. Nor were these reasons anything a modern doctor would recognize: for instance, the Anglo-Saxon scholar known as The Venerable Bede (c. 672–735) attributed certain diseases to bad luck or darts shot by elves.

At the time, evil spirits were also commonly thought to cause disease. As late as the 1500s, people believed that demons invaded the bodies of those who were physically or emotionally weak, causing a range of ailments from sore throats to pneumonia to mental illness. Supposedly these disease-causing demons traveled with winds from the north, so many people covered their faces or stayed indoors to avoid north winds.

Popular remedies

So-called remedies for illnesses included drinking potions made from water in which frogs had been boiled or taking walks on a moonlit night. Charms or amulets, objects believed to possess special powers, were another popular cure. These superstitions are still reflected in the custom of wearing heart-shaped lockets around the neck—a tradition based on the medieval idea that the heart possesses spiritual powers.

People often put garlic or herbs in the locket to ward off diseases, witches, or demons, which in the view of medieval Europeans were different aspects of the same thing. Even as late as the sixteenth century, England's Queen Elizabeth I (1533–1603)—unquestionably one of the most sophisticated people of her time—wore an engraved gold ring suspended from her neck to drive away "bad airs."

Life Science

ESSAYS

Blood-letting by a Dutch barber-surgeon. A common medieval medical practice, blood-letting was believed to restore the balance of body fluids. (Illustration by Corne Dusart. Bettmann/Corbis. Reproduced by permission of the Corbis Corporation.)

Health-care "professionals"

Lacking doctors trained in scientific medical care, people in the Middle Ages and Renaissance depended on barber-surgeons, apothecaries, and herbalists. Few European villages had their own physician, and therefore the barber served a dual function, performing medical work with the same razor he used for trimming hair and beards.

Barbers also set broken bones and pulled teeth, but primarily they performed surgery, including bloodletting. In this gruesome procedure, a

Life Science

ESSAYS

patient gripped a staff or a stick in order to make his or her veins stand then the barber made his incision or cut. What followed was a scene of almost unimaginable misery. It is a miracle than anyone survived the brutalities of medieval medicine. But many did, and after completing a "successful" surgery, the barber-surgeon attached the bloody bandages to the staff to dry. Often wind would blow the bandages around the staff, and eventually this staff became the symbol of barbers, reflected in the red-and-white barber's poles still used today.

Apothecaries

The apothecary, who prepared and sold medicines, was the forerunner of the modern pharmacist, but apothecaries were more like witch doctors or fortune-tellers; they relied on experience, superstition, and folklore rather than on scientific methods for curing people. An example of the "medicine" sold in apothecary shops was mumia, a powder derived from dried Egyptian mummies. The fact that it supposedly possessed the power to cure diseases created a great demand for mumia, which in reality came from the remains of recently executed prisoners.

People also relied on powder supposedly ground from a unicorn's horn, believed to cure any poisoning and bring good fortune. When French surgeon Ambroise Paré (1510–1590) suggested that many apothecaries were growing rich substituting domestic animal horns for those of the imaginary unicorn, the dean of the Paris Medical College denounced him.

Herbalists

As with the apothecary, the work of the herbalist has a modern counterpart, in this case in the herbal cures that are a part of today's alternative medicine (see essay "'Alternative' Medicine Begins in the East" in this chapter). Herbalists in premodern times were famous for dispensing potions of dried herbs and plants, parts of dead animals, and sometimes even excrement or dung. The herbalist's approach, however, was trial and error: the herbalist would tout a particular potion as a cure, whereupon a sick or injured person bought it. If the person actually got better, the herbalist recorded the herbal potion as successful and added the concoction to her medicinal collection. Occasionally, an herbalist stumbled across a plant with genuine curative powers. In 1638 the Countess of Chinchon, wife of the Spanish Viceroy of Peru, was cured of malaria by an extract of the bark of the Peruvian quinaquina tree. The news of the cure spread quickly throughout Europe, and the drug was named *chinchona* in honor of the countess. Now known as quinine, the drug remains an effective weapon in the treatment of malaria. More often, herbalists and even apothecaries stocked medicines that had little effect on the diseases for which they were intended.

Life Science

ESSAYS

Powder made from the horn of a unicorn, an imaginary animal, was believed to cure any poisoning and to bring good fortune. (©Araldo de Luca/Corbis. Reproduced by permission of Corbis Corporation.)

Saints, relics, and pilgrimages

In such an environment, it is not surprising that many people were searching for alternatives, including praying to Christian saints and heroes of the past. Each saint became associated with particular aspects of life, and were thought to offer cures for specific diseases. These associations were often based on fanciful tales concerning the saint's life.

For instance, it was said that Saint Margaret of Antioch had been swallowed whole by a dragon, but when she made the sign of the cross, the cross grew, bursting open the dragon and releasing her. She therefore became the patron saint of pregnant women. Desperate for healing, people prayed to the saints, and many collected relics or holy objects associated with them. For instance, the chains that had held the apostles Peter and Paul captive under the Roman emperor Nero were said to have the ability to cure blindness.

Life Science

ESSAYS

*OPPOSITE PAGE
An illustration of Thomas Becket being attacked by soldiers while a monk watches. The Catholic church declared Becket a saint shortly after his death and people reported miraculous cures from touching his bloodstained garments.
(© Leonard de Selva/Corbis-Bettmann. Reproduced by permission of the Corbis Corporation.)*

People also embarked on pilgrimages, visiting places associated with events in the Bible or with the lives of one saint or another in an attempt for help. In England the most popular pilgrimage destination was Canterbury, where the bishop Thomas Becket (c. 1118–1170) was murdered. Becket was assassinated by order of King Henry II of England (ruled 1154–1189) because Becket opposed the king's attempts to gain power at the expense of the church. Shortly after his death, the church recognized Becket as a saint, by which time hundreds of people had reported miraculous cures for blindness and other conditions simply as a result of touching the bloodstained garments he had worn at the time of his murder. The pilgrimage to Canterbury had long since become a tradition by the time English writer Geoffrey Chaucer (c. 1342–1400) based his *Canterbury Tales* around it.

The Crusades

The most important pilgrimage destination in Europe was Rome, headquarters of the Catholic church; but the greatest aim of the medieval pilgrim was Jerusalem, the city where Jesus Christ, founder of Christianity, had been crucified. Since the 600s, Jerusalem and the rest of the Holy Land (the area equivalent to modern-day Israel and Lebanon) had been under the control of Muslims. The Muslims allowed Christian visitors to come and go freely, but in 1095 the pope and other European leaders claimed that God had called on them to win the Holy Land for Christianity. The result was the Crusades, a series of wars lasting nearly two centuries from 1095 to 1291.

Though the Crusades were justified at the time as holy wars, and later romanticized with tales of brave knights fighting for the cross, most crusaders were motivated by the desire for riches and adventure, or even the thrill of shedding blood. The Crusades themselves, after a few initial successes, were a military disaster, and they stirred up tensions between Christians, Jews, and Muslims that still persist today. Yet for all the viciousness behind them, the Crusades are considered the greatest thing that ever happened to medieval Europe.

Changing attitudes

By coming in contact with Muslims, Europeans had their first exposure to a truly vibrant, scientifically advanced society (see essay "The Great Doctors of Medieval Islam" in this chapter). In the years that followed, medical texts from the Islamic world, many of them based on works from ancient Greece, appeared as translations in Europe, where they educated generations in the health sciences. Medical schools flourished and scientific medicine gained the support of powerful figures such as Holy Roman

Life Science

ESSAYS

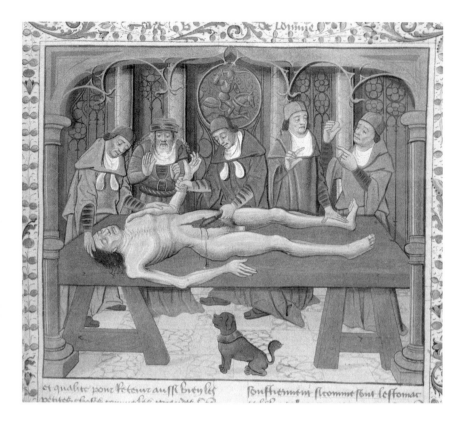

A depiction of a dissection from a Renaissance anatomical manuscript. In 1224 Holy Roman Emperor Frederick II was the first major European ruler to permit dissection of cadavers which had been prohibited since the approximately the fifth century B.C.E. (© Archivo Iconografico, S.A./Corbis. Reproduced by permission of the Corbis Corporation.)

Emperor Frederick II (1194–1250), who in 1224 issued the first laws regulating the study of medicine. In 1241 he became the first major European ruler to permit dissection of cadavers (that is, cutting open dead bodies for medical study), which had been prohibited for spiritual reasons from the time of the Greeks.

Beginning in about 1550, several factors began to elevate the status of scientifically trained physicians across Europe. The Renaissance, by then long underway, was characterized by a heightened interest in the classics of Greek and Roman civilization, and these reintroduced the best aspects of ancient medicine, including careful observations of patients and their symptoms. French philosopher **René Descartes** (1596–1650; see biography in Math chapter) introduced the concept of the body as a mechanism or machine rather than something mystical and unknowable, while anatomists such as **Andreas Vesalius** (1514–1564; see biography in this chapter) began to investigate the mysteries of the body firsthand. The invention of the printing press in 1450 by **Johannes Gutenberg** (c. 1395-1468; see biography in Technology chapter) in Germany made possible the spread of medical literature and the sharing of information across Europe.

Lithotomists and Other Wandering Quacks

Because western Europe in the Middle Ages and the Renaissance had few laws governing surgery, untrained doctors wandered the countryside, filling marketplaces and county fairs. Among these "doctors" were tooth-pullers and cataract-removers (a cataract is a clouding of the eye's lens). With no anesthesia, no concept of an antiseptic (germ-free) operating environment, and no antibiotics with which to treat the patient afterward, surgery was a ghastly affair. Yet people needed cures for their ailments and were willing to undergo almost anything to get them.

Another type of doctor offering services at the time was a lithotomist (li-THAHT-uh-mist), or "cutters-of-the-stone," who specialized in the quick surgical removal of bladder stones. Bladder stones were a common problem due to untreated urinary tract infections, and they posed a serious threat by potentially blocking the passage of urine. With so many individuals, particularly males, in need of treatment and therefore willing to pay for it, many lithotomists made a handsome living in premodern Europe. Among them was the notorious Jacques de Beaulieu (1651–1719). After an apprenticeship to a wandering Italian surgeon in 1690, Beaulieu donned monk's robes and called himself Frère Jacques—believed to be the source of the nursery rhyme by that name—and was eventually expelled from France due to the number of patients who died from his lithotomy techniques.

Separating religion and science

One of the key aspects of Europe's reawakening lay in achieving something even the ancients had never managed to do: the distinction between beliefs based on things that people cannot see or touch, and those based on things that can be observed and measured. Science, of course, is entirely about judgments formed from observation, not faith; whereas both religion and superstition, though not mutually exclusive, revolve around believing in something that cannot be seen.

There is nothing wrong with the application of religious faith in a medical situation, but most of the world's faiths believe that this must be done in conjunction with the treatment of a properly trained and certified

Life Science

ESSAYS

physician. Indeed, many medical experts today maintain that prayer aids the process of healing by placing a positive focus on a patient's thoughts.

Christianity and science
Certainly there are aspects of Christianity which, when emphasized at the expense of scientific reasoning, have unfortunate results. For instance, Christianity stresses the idea that God has a purpose for everything that happens, and when taken to its extreme in the Middle Ages, this meant that a person could not do anything to stop misfortunes such as diseases, which were seen as God's punishment for sin. In addition, Christianity holds that the body and the spirit are separate, and because it places a higher value on spiritual matters than on physical ones, many people took this to mean that physical concerns were unimportant.

However, to blame Christianity for the general lack of progress in the Middle Ages is to oversimplify the facts. With the decline and fall of the Roman Empire, the Christian church was the only strong institution remaining in western Europe, and without its influence, the disorder that characterized the period would have been much worse. Furthermore, the church encouraged learning, and most scholars in the Middle Ages were priests or monks. Separated from the rest of society in monasteries, Christian monks opened their doors to the sick and the needy. Indeed, if any reliable medical care was to be obtained in medieval Europe, it was either at a monastery or a hospital established by the church.

☐ THE REAWAKENING OF THE LIFE SCIENCES IN EUROPE

Over a period from about 1100 to 1500, Europe experienced mammoth changes that affected virtually every aspect of life. Though the Crusades (1095–1291), a series of religious wars in which Europeans attempted to gain control over the Muslims who ruled the Middle East, proved a military failure, resulting in no net gains of lands in the Middle East, they brought about a reawakening of intellectual life, which had stagnated for about a thousand years.

When the Crusades began, the pope, as spiritual and political leader of the Catholic Church, was the most powerful figure in Europe; by the end, however, leaders of independent states increasingly challenged his political authority. The 1300s saw the rise of the concept of nationhood among Europeans, who had previously been content to see themselves primarily as Catholics, and only secondarily as English or French.

The emerging sense of national identity brought conflicts between rulers and the pope, and eventually these challenges to old-fashioned

Words to Know

anatomy: The study of the structure of organisms, including the human body.

contagion: The transmission of a disease by direct or indirect contact.

dissection: Cutting open dead bodies in order to study their internal characteristics.

epidemic: A disease that affects large numbers of people.

public health: A set of policies and methods for protecting and improving the health of a community through efforts that include disease prevention, health education, and sanitation.

quarantine: The isolation or separation of people or items in order to prevent the spread of communicable diseases.

authority spread to other sectors of life as well. Just as rulers vied for power with the pope (and with each other), so the nobility or aristocracy of each nation tried to gain influence at the expense of their rulers. Meanwhile the reemergence of trade, which had been almost purely local during most of the Middle Ages (roughly 500 to 1500), brought into existence a middle class who likewise fought for political influence.

With kings threatening the power of popes, nobles challenging the authority of kings, and the middle classes increasingly vying for a place alongside the nobility, naturally scholars and thinkers began to challenge the intellectual authority of the church. These thinkers also began to reject old ideas and old ways of doing things, including concepts held over from the ancient period, which by then was already fifteen hundred years or more in the past.

It was a time of great change accompanied by great fear. As the Crusades came to an end, Europe faced the threat of invasion by the Mongols, a nomadic tribe from central Asia that conquered much of the known world. The Mongols never gained a foothold in Europe beyond Russia, however, and by establishing a vast empire they stimulated the resumption

Life Science

ESSAYS

of intercontinental trade and the beginnings of European exploration. For the first time in centuries, people could come and go with relative ease between East and West. Though this had its obvious advantages, it also made possible the spread of the Black Death (1347–51), an epidemic that killed almost one-third of Europe's population.

The rise of universities
Until about 1100, physicians learned their trade in much the same way as carpenters or bricklayers, in an apprenticeship to a trained master. Like tradespeople, they organized into guilds, an early type of labor union that emerged in the 1100s. All of that, however, would change with the rise of university education that occurred in the century that followed. The rediscovery of ancient Greek and Roman texts, preserved during the Middle Ages by Muslim scholars, greatly influenced these events.

The first European medical school, established in the 800s at Salerno in southern Italy, had become an influential center of learning by the eleventh century. At that time most of Italy and what is now Germany fell under the control of the Holy Roman Empire. In 1221 the emperor of the Holy Roman Empire, Frederick II (1194–1250) ordered that no doctor could legally practice in the empire unless he had passed an examination by the authorities at Salerno, a move that greatly increased the status of trained physicians.

By then, a number of other medical schools had sprung up in towns such as Bologna, Italy, as well as in Paris, France, and other major cities. This movement corresponded with the rise of universities, large institutions of higher learning that offered an education in a variety of arts and sciences. The result was a growing population of trained and certified medical professionals, who gradually began to replace the barber-surgeons and various quacks (untrained medical practitioners who relied on unproven methods, folklore, and superstition rather than scientific methods for healing) who had dominated European medicine throughout much of the Middle Ages.

Developments in surgery
Bologna became the home to new advances in surgery under the leadership of various teachers, including Hugh of Lucca (c. 1160–c. 1257); his student, Theodoric (1210–1298), Bishop of Cervia, who recorded Hugh's ideas in a book called *Chirurgia* (Surgery); and Mondino dei Liucci (c. 1270–c. 1326). Mondino was particularly important because he taught anatomy through the practice of public dissection. For many thousands of years, dissection had been prohibited because people believed it was an offense to God or the gods to cut open a human body, but it was essential if physicians were to learn how the body works.

Italy was not the only center of medical learning during the late Middle Ages; in fact, during the 1300s the focus shifted to France, whose most notable surgeons included Henri de Mondeville (1260–1320) and his student Guy de Chauliac (c. 1300–1368). Mondeville stressed the importance of studying anatomy, an emphasis reflected in de Chauliac's monumental work, *Chirurgica magna* (Grand surgery). When the Black Death, a form of the plague, struck the French town of Avignon, Guy chose to stay on rather than flee to the countryside, as many people were doing. Though infected by the plague, he survived and wrote a valuable medical account of the disease and its spread.

Also significant was John of Arderne (1306–1390), the first important English surgeon, who operated on wounded soldiers during the Hundred Years' War (1337–1453). A conflict between England and France, the Hundred Years' War, whose most well-known participant was Joan of Arc (c. 1412–1431), was the first major military action in which firearms were used.

Life Science

ESSAYS

The plague

Despite its length, the Hundred Years' War resulted in only a fraction of the deaths caused in just five years by the plague, an epidemic from 1347 to 1351 that killed almost one-third of Europe's population. The plague was caused by a bacteria found in certain rat fleas and was carried by rodents on ships to the ports of Europe. The plague that struck Europe during this epidemic was the bubonic plague. Also referred to as the Black Death, this is just one of several forms that the plague can take.

In the bubonic variety, people's lymph nodes—glands that normally protect against infection—swell and become hard before turning black and bursting. At least half the people who contracted this form of the plague died.

The bubonic plague spread throughout Italy in 1347, and by 1348 it had reached Paris, France, where by the end of the year eight hundred people a day were dying. It rapidly infected the entire European continent, and by 1351 sailors arriving in Greenland found its ports deserted. Though the worst phase of the epidemic ended in 1351, it continued to spread, reaching Moscow, Russia, by 1353, and the next five centuries saw occasional outbreaks of the disease. As late as 1894, a strain of plague killed over six million people in Asia over the course of fourteen years.

Impact of the plague

Not only did the plague kill some thirty million people, but it also wreaked havoc on the very fabric of society. Desperate with fear, people turned against one another to protect themselves. Sometimes parents even

Life Science

ESSAYS

abandoned their children, and plague victims became such outcasts that some were even walled up inside their houses and left to die.

Lacking a scientific understanding of disease, many people saw the plague as God's punishment for sin. In an appeal for divine mercy, many prayed to saints or made pilgrimages, while an extremist group known as the flagellants publicly whipped themselves as a way of making atonement. One of the most sinister outgrowths of the plague was a wave of violence against Jews, who were rumored to have started the epidemic by poisoning wells. In addition to the many Jews who died in the plague itself, many more perished in mass executions.

For all its negative impact, the plague also had a number of surprisingly positive effects. By wiping out a large population of workers, it created a great demand for labor, meaning that the peasants and skilled tradespeople who survived could charge more for their services. This led to an overall increase of living standards. On a deeper level, the plague led people to question the authority of the pope and other church leaders who claimed to speak for God. These leaders had not been able to predict the destruction that swept Europe, nor were they able to offer any relief from it. In fact, at the time the plague hit, the popes were caught in an intense political struggle with Europe's kings, and they seemed to have more interest in their own power than in the needs of the people. Disillusionment with the popes led to increased questioning of old ideas and opened people's minds to new ones. The ultimate outcome of these changes would be the Renaissance, a rebirth of the arts and sciences that lasted approximately from 1350 C.E. to 1600 C.E.; and the Reformation, a revolt against the authority of the Catholic Church that led to the establishment of Protestant denominations.

Not all of the reactions to the plague were as superstitious. Physicians also looked for scientific explanations and physical causes for the plague. For perhaps the first time, doctors began to realize that disease was not an

Townspeople leaving a city to avoid the plague. Between 1347 and 1351 the bubonic plague killed almost one-third of Europe's population. (Reproduced by permission of the Mansell Collection/Time Inc.)

Science, Technology, and Society

individual matter, the result of imbalances in a patient's "humors," or bodily fluids. This had been the accepted wisdom throughout the Middle Ages, but the extensive outbreak of the plague suggested to doctors that disease could be spread from person to person. This realization led to the first theories of contagion, the idea that diseases could be transmitted by direct or indirect contact. It would be many centuries before the development of germ theory in the 1800s, but already by 1546 the Italian physician Girolamo Fracastoro (c. 1478–1553) was suggesting that illnesses spread through the transmission of disease-carrying "seeds."

A new concept of public health

Another consequence of the Black Death was the return, for the first time since ancient Rome, of the concept of public health. Public health refers to a set of policies and methods for protecting and improving the health of the community through efforts that include disease prevention, health education, and sanitation.

The Romans fostered public health through the application of engineering, developing large sewage and drainage systems, as well as aqueducts or pipes that brought fresh water to cities. However, the decline and fall of the Roman Empire over a period from about 200 to 500 C.E. brought a virtual end to the concept of public health for about a millennium.

Quarantines and sanitary laws

With the Black Death and the ideas of contagion that followed, towns began to take consistent measures to separate not only the sick but—and this was a key improvement—people who might be sick or might be carrying some disease. Late in the 1300s, officials in the Italian port of Ragusa began requiring arriving ships to wait at sea for a period of forty days in order to confirm the health of the crew. They called this policy *quaranti giorni*, or forty days, a term from which the word *quarantine* is derived.

In England, King Richard II (1367–1400) established the first national "sanitary laws" intended to combat the spread of disease, and soon a number of towns adopted similar measures. These laws created a legal system of sanitary control, using observation stations, isolation hospitals, and disinfecting procedures. In addition, towns improved sanitation by developing methods of obtaining pure water supplies, inspecting food, and disposing of garbage and sewage.

The new ideas of Paracelsus

All over Europe a more complete understanding of disease was emerging, and though a number of doctors and scientists contributed to these new ideas, few had as great an impact as the Swiss physician who went by the

Life Science
ESSAYS

name of **Paracelsus** (1493–1541; see biography in this chapter). With his combination of ideas, some modern and others medieval, Paracelsus was a man with his foot in two worlds.

Influenced by alchemy, a medieval pseudoscience that involved attempts to turn ordinary metals into gold, Paracelsus strongly advocated the use of chemically prepared medicines to treat illness. He identified diseases as specific medical entities that attacked particular parts of the body, a concept at odds with the old medieval doctrine of the four humors, the belief that human health is governed by a balance of four bodily fluids: blood, phlegm, yellow bile, and black bile. But Paracelsus was never afraid of controversy. He made a name for himself by condemning the way that medicine was taught in European universities, and in the process he gained a number of enemies among the medical establishment.

Though the establishment and development of universities and medical schools had enabled the spread of medical education, by Paracelsus's time their methods of teaching had become stagnant and old-fashioned. Professors still taught from ancient books of medicine, full of misconceptions such as humoral theory, and many looked down on the idea of conducting dissections (cutting open dead bodies for medical study) themselves. Instead, they oversaw the dissections, actually performed by assistants, while reading aloud from medical texts.

Continued epidemics such as the outbreak of syphilis, a sexually transmitted disease that causes painful aches and fevers, disfiguring sores, and often death, had led to a growing public dissatisfaction with established medical techniques. Paracelsus was at the forefront of this challenge. Whereas old-fashioned physicians recommended the use of herbal remedies and changes in diet, Paracelsus advocated strong remedies, often containing metallic elements. The old method had been to "cure by contraries": if someone seemed to have too much of one humor, the doctor would prescribe something to counteract its influence. Paracelsus, however, maintained that "like cured like," meaning that the physician had to identify a similarity between the disease in the body and the substance in the outside world that could be used to cure it. His ideas met fierce challenges at the time, but over the

Gerolamo Fracastoro suggested that illness was spread through disease-carrying "seeds" in the sixteenth century, many centuries before the development of germ theory. (Reproduced by permission of the Granger Collection, New York.)

centuries that followed, they would become an established part of medical practice.

Life Science

ESSAYS

▢ PRINTING, ILLUSTRATION, AND DISSECTION

The period from about 1350 to 1600 is known as the Renaissance, a time of rebirth in the arts and sciences. The Renaissance also coincided with a movement called the Reformation, in which religious leaders throughout Europe questioned the authority of the pope and the priests of the Catholic Church as the sole interpreters of the Bible.

Both movements represented a move into the future, yet each also involved a rediscovery of the past. Leaders of the Reformation encouraged Christians to read the Bible themselves and not to use the church as an intermediary. As for the writers and artists of the Renaissance, they turned to ancient Greece and Rome, societies in which learning had flourished.

To an extent, priests and scholars in the medieval period (c. 500–c. 1500) had kept alive the ideas passed down from the Bible, as well as those of the Greeks and Romans, yet they had done so in a very rigid fashion. In the health sciences, for instance, doctors had made little effort to improve on what they had learned from the ancient Greeks about medicine. Medical minds of the Renaissance, however, sought to find what was good in the ideas of Greek physicians, and to challenge what was not.

The invention of the printing press in 1450 spurred on these changes in attitude. Until then, learning had been entirely in the hands of the church, whose priests communicated primarily in Latin. The printing press, however, made the written word available to everyone. Not only did this encourage education and the spread of national languages such as English, French, German, and Italian, but it also led to the dissemination of ideas by leading figures of the Renaissance and Reformation. Particularly important for the study of health sciences was the new availability of illustrated medical texts, which made use of information gathered by the dissection of human bodies.

The spread of printed medical books

Until the invention of the printing press, scholars in the health sciences and other areas had to rely on medical books that were laboriously handwritten and recopied through the centuries. Not only was this method of distributing medical information very slow, but it meant that books were available to only a select few. Also, the quality of those books diminished over the centuries, as frequent changes to the content and illustrations led to increasing inaccuracies.

The French Disease

"The Syphilitic" woodcut by Albrecht Durer.

In 1494, French troops attacked the Italian city of Naples, beginning a sixty-five-year series of conflicts loosely named the Italian wars, during which French kings, Holy Roman emperors, and popes vied for control of Italy. Returning from that first engagement, the French brought home a strange new ailment—the first major sexually transmitted disease. Initially dubbed *morbus gallicus,* or the "French disease," it soon spread throughout Europe, bringing with it painful aches and fevers, disfiguring sores, and often death. In 1530 the Italian physician and humanist Girolamo Fracastoro (c. 1478–1553) gave it a new name in a poem about a young shepherd boy infected with a "pestilence unknown." The shepherd's name was Syphilis.

Actually, there are four kinds of syphilis, only one of which is transmitted through sexual contact. In addition to sexual transmission, venereal syphilis can affect an unborn baby whose mother has the disease. Syphilis

Early texts had been written on heavy parchment, which was made from the skin of a sheep or a goat. By the 1200s paper, invented in China centuries before, had become more widely available, and this facilitated the production of books even before the printing press. Doctors who had copies of the most popular medical texts of the day, written by Greek, Arab, and European physicians, would add their own observations and corrections in the notes. Still, there was no way of creating many copies of identical texts for wide distribution.

Soon after the development of the mechanical printing press by **Johannes Gutenberg** (c. 1395–1468; see biography and essay "The Printed

is characterized by lesions or sores on the genitalia (organs of the reproductive system). Over time the lesions can spread to affect other parts of the body—even the brain. Today, doctors recognize the bacteria that causes syphilis and can treat it with antibiotics; therefore, great strides have been made in treating the disease.

Concepts such as bacteria and antibiotics, however, lay centuries in the future when Europe experienced its first outbreak of syphilis. Most of the treatments used at the time—bloodletting, for instance, or ingestion of mercury (a heavy silver-white poisonous metallic element)—caused more harm than good. Because they could not find a cure, physicians directed their attentions to arguing whether the disease was a new one or not. Many believed that it had come from the New World and been brought back to Europe by sailors under Spanish explorer Christopher Columbus (1451–1506). This is quite possible, though in fact Europeans took far more diseases, including smallpox, *to* the New World, and these had a devastating effect on the native populations of the Americas.

As with the Black Death, the spread of syphilis influenced new ideas about disease. Whereas physicians in the Middle Ages had believed that illnesses came purely from within the patient's body, as a result of imbalances in the "humors," doctors such as Fracastoro suggested a new explanation: contagion, or the idea that diseases could be spread from person to person.

Word" in the Technology and Invention chapter) in Germany, books went into rapid production. One of the most important changes for the medical profession, aside from the obvious fact that textbooks became more widely available, was the addition of more and better illustrations. As long as people were copying books by hand, the quality and the quantity of the illustrations depended on the abilities of the person who copied them; now a machine could create an exact duplicate of an original illustration.

The invention of the mechanical printing press brought knowledge of medicine to the populace. By making possible the spread of texts in various national languages rather than in Latin, the printing press further

Life Science

ESSAYS

enabled people who had not been trained in universities to learn about new developments in the medical arts.

Changing themes in art and illustration

One of the leading components of the Renaissance was a change in the visual arts. Up until that time, artists had been faceless figures who created whatever the pope or the local bishop commissioned for a church or other religious building. No one knows the names of those masters who fashioned the magnificent sculptures of medieval cathedrals, for instance, but the leading artists of the Renaissance—such as **Leonardo da Vinci** (1452–1519; see biography in Technology and Invention chapter)—became so widely renowned that their names are still household words.

At the beginning of the Renaissance, painters and sculptors dealt almost exclusively with biblical themes. The portrayal of events from the Old Testament and the New Testament (for instance, Leonardo's *Last Supper*) would continue throughout the Renaissance, but artists also began to branch out into new subjects: depicting the classical world of Greece and Rome, for instance, or even the everyday life of Renaissance Europe. Indeed, Leonardo's most famous work, the *Mona Lisa,* portrays neither a figure from Greece, Rome, *nor* the Bible—rather, she was apparently a wealthy Italian woman.

Even during the Middle Ages (roughly 500 to 1500), it had been necessary from time to time for illustrators to depict scenes other than those in the Bible—for instance, for the purpose of commemorating historical events. Likewise, medical books needed illustrations, but mostly these were vague, general sketches without much detail. An example was the "zodiac man," a common type of illustration showing the points for bloodletting that corresponded to the astrological signs. Other medieval medical illustrations were simply grotesque, possessing little value in the way of medical education. For instance, a picture might show the grinning figure of Death standing by the bedside of a dying patient.

Depicting the body

In rediscovering ancient Greece, Renaissance scholars likewise rediscovered an object of amazing complexity and beauty that had been, quite literally, under their noses all along: the human body. The Bible emphasized spiritual matters over physical ones, and medieval Christian leaders had taken this to an extreme, treating the body as something insignificant and unworthy of attention. By contrast, the Greeks—whose enthusiasm for athletics was reflected in their Olympic Games—admired the human body and took joy in the muscular forms of men and the shapely curves of women.

Life Science

ESSAYS

"David" by Michelangelo. Michelangelo's sculpture reflects a trend toward more accurate depictions of the human body during the Renaissance. (Reproduced by permission of Scala/Art Resource, New York.)

Life Science
ESSAYS

This "Greek" attitude toward the body is reflected in an engraving called *The Battle of Ten Naked Men* by the Italian artist Antonio Pollaiuolo (c. 1431–1498). The scene depicts a group of men in combat, bristling with so many exterior muscles that the result is actually unrealistic. Nonetheless, works such as Pollaiuolo's reflected a growing interest in the human form, and their inaccuracies made artists aware of the need for further study in anatomy. In order to achieve this, they would have to witness the dissections of human bodies by physicians.

An ancient Roman copy of Discobolus by Myron. The form of the discus thrower displays the Greeks' mastery of human anatomy. (© Gianni Dagli Orti/Corbis. Reproduced by permission of the Corbis Corporation.)

Mondino pioneers human dissection

Though the Greeks had admired the living human body, religious and social customs had prevented Greek physicians from actually cutting open dead humans. Tampering with the body, they believed, was the same as tampering with the soul, and for this reason, early physicians such as **Galen** (129–c. 199; see biography in this chapter) had to base their anatomical information on dissections of animals. This, of course, resulted in a number of incorrect conclusions when applied to humans.

The prohibition against dissection continued throughout most of the Middle Ages because the Catholic Church maintained that a mutilated body could not be resurrected to experience the afterlife in heaven. Thus Mondino dei Liucci (c. 1270–c. 1326), the first significant European physician to dissect humans instead of animals, must have been a very brave individual indeed.

Mondino worked in the medical school at Bologna, Italy, a highly respected institution (see the essay "The Reawakening of the Life Sciences in Europe" in this chapter); nonetheless, his act of cutting into the first cadaver (dead body) defied established morals. Yet thanks to Mondino, who also wrote the first textbook on anatomy—a highly influential text for the next two centuries—dissection of humans became an accepted practice. In time, it would become an essential component of training for medical students and later for Renaissance artists.

Leonardo and Dürer

Leading the movement to combine art and science in precise depictions of the human body was Leonardo da Vinci. His notebooks included visual studies of the arm and forehand, motor muscles of the hands, and the actions of the muscles in breathing. He also spent hours personally dissecting and studying corpses, from which he created some 750 anatomical drawings. In 1489 he began work on an incredibly ambitious project, an anatomical atlas showing the stages of human development from birth to death, though he never completed it.

One of the techniques that Leonardo pioneered was that of rotation, or showing the same body part from a number of different angles. Though this might seem obvious now, it was revolutionary at a time when even the concept of perspective (showing distant objects smaller than close ones) was new. He also introduced the idea of transparency—in other words, treating exterior tissue as though it were invisible, so that he could represent interior bones and organs.

Leonardo's anatomical drawings and their accompanying text were not published until the late nineteenth century, yet they had an impact in his own time because other artists saw his notebooks and copied his work. Among the contemporaries he seems to have influenced was the great German artist Albrecht Dürer (1471–1528). Most famous for *Praying Hands,* Dürer produced incredibly accurate representations of sinews (muscles) and veins on the human body, but he also created numerous drawings and paintings of plant and animal life.

Dürer's Turf *influences botanical illustration*
Of the many great works produced by Dürer, none was more important in terms of its influence on botanical illustration than *The Great Piece of Turf* (1503). A life-sized depiction of grasses and dandelions, it represented a shocking departure from the tradition, still strong even in the Renaissance, of representing "important" subjects such as those of the Bible and Greek history or mythology. Dürer, by contrast, had chosen a simple scene such as one might find at the edge of any pond, yet his painting had a tremendous artistic and scientific impact in Europe.

Up to that time, the majority of botanical texts had been herbals, medieval plant encyclopedias that focused on the medicinal uses of various herbs. But with the changes taking place in the life sciences—particularly due to the introduction of plant species from the New World—there was a demand for books that dealt more scientifically with the structure and the classification of plants. Dürer's *Turf* provided a model for the detailed, exceedingly accurate illustrations that such books required.

Publishers had previously been content to keep using the same old, inaccurate woodcuts (ink engravings made by cutting a picture into a block of wood) from medieval herbals. This was cheaper, of course, than

Life Science

ESSAYS

Life Science
ESSAYS

hiring artists to produce new, accurate illustrations. Dürer, however, raised the standards, and from about 1500 to about 1700 much higher quality botanical illustrations were being created.

New medical and botanical illustrations

The work of artists such as Leonardo and Dürer had an enormous, if sometimes indirect, impact on changes in the portrayal of human bodies and other subjects that accompanied new texts in the life sciences. One remarkable contemporary of these artists, a man almost certainly aware of their work, was the Italian anatomist and surgeon Giacomo Berengario da Carpi (c. 1460–1530), who illustrated his own writings.

Like Leonardo, Berengario da Carpi depicted bodies with skin removed and provided multiple views of a single body part. He often portrayed skeletons or partly dissected cadavers as though they were standing and observing a landscape. This feature, too, seems to suggest the influence of trends in Renaissance art, which was noted for its sometimes odd use of natural settings—for instance, the rugged backdrop to the *Mona Lisa*.

In Berengario da Carpi's illustrations, for instance, a figure stripped of his skin might be shown leaning against an ax, observing the hills around Bologna. In another picture, a skeleton stands in front of a grave, from which he had obviously been taken, holding two other skulls in his uplifted hands.

In the 1530s German physician Otto Brunfels (c. 1488–1534) published *Herbarum vivae eicones* (Living images of plants), a three-volume botanical work whose most significant feature was not Brunfels's text, but the illustrations of Hans Weiditz. As with Dürer's *Great Piece of Turf*, it was clear that Weiditz had used real specimens for his models, even including leaves partially eaten by insects.

In fact Weiditz's approach may have been a bit *too* artful, in the sense that his pictures were filled with distracting details. By contrast, the five hundred illustrations, made by three artists, that accompanied *De historia stirpium* by German physician and botanist Leonhard Fuchs (1501–1566) showed ideal plants with no flaws, so that the viewer's attention remained

"Mona Lisa" by Leonardo Da Vinci. Renaissance art is noted for its sometimes odd use of natural settings, as seen in the rugged background of the Mona Lisa. Medical illustrators at the time also sometimes used in such natural settings. (Reproduced courtesy of the New York Public Library Picture Collection.)

Public Dissections and Grave Robbing

Once forbidden, dissection had become a valued aspect of medical training by the late 1500s, a fact that carried with it some unintended—and gruesome—consequences. Because dissections were held in public, this meant that not only serious observers such as medical students and artists could attend: anyone else who simply wanted to see a corpse cut open could watch as well.

Eventually dissections became a form of entertainment, announced ahead of time as though the impending event were a play. And indeed the public dissections of the late Renaissance had the atmosphere of a play, complete with curtains for the "stage." Youngsters even sold refreshments.

Initially medical schools got their corpses from criminals who donated their bodies prior to execution. However, with so many medical students eager to get their hands on a corpse, there were simply not enough executed prisoners to go around, and this led to the grim practice of grave robbing. Students and professors would loot graves from nearby towns, assuming that no one participating in the dissection would know the deceased. This only applied with regard to a freshly dead body, of course; skeletons, by contrast, were less identifiable, and therefore relatively easy to steal.

focused on the plants themselves rather than the artistry with which they were presented. Subsequent botanical artists used this approach, because the purpose of the illustrations was to represent the species as accurately as possible. This trend continued with the work of German naturalist Maria Sibylla Merian (1647–1717), one of the few female writers and illustrators of texts in the life sciences during the period.

The illustrated Vesalius

One of the greatest figures in the life sciences during the late Renaissance was **Andreas Vesalius** (1514–1564; see biography in this chapter), whose *De humani corporis fabrica* (On the structure of the human body) benefited enormously from the inclusion of 663 illustrations. The artist who created these illustrations apparently had a sense of humor: for the frontispiece, the picture facing the title page, he depicted Vesalius performing a dissection before his students, while a group of discredited barber-surgeons crouch under the table and a monkey and dog compete to get into the picture.

The identity of that artist has been a subject of debate. Some scholars believed that it was the great Italian painter Titian (c. 1488–1576), though more likely the illustrator was Vesalius's Dutch assistant, Johannes Stephanus of Calcar. What is not debatable is the value of the illustrations themselves or of the text that they accompanied.

Vesalius had grown up a great admirer of Greek physician Galen (129–c. 199), whose ideas still dominated European medicine centuries after his death; but after Vesalius began studying anatomy firsthand, he became one of Galen's fiercest critics. By promoting an understanding of the structures in the human body, *De humani corporis fabrica* set in motion great changes in the world of medicine. Thanks to Vesalius, doctors would increasingly seek understanding by studying the body, not by studying outdated medical texts.

☐ THE MEDICAL DISCOVERY OF WOMEN

Throughout history more than half of the people involved in health care and healing have been women. Women have also made up more than half of the patients. Historically, the disproportionate fame and recognition given to male practitioners is due largely to the fact that surviving manuscripts were written by men and because women generally were not accepted into medical schools. Women have long practiced medicine, but in premodern times dealt primarily with childbirth and conditions of the female reproductive system—women took care of women.

During the Renaissance, a period of artistic and intellectual rebirth that marks the end of the Middle Ages, the use of herbs and potions for healing led to associations between healing and witchcraft, in some cases resulting in unjust trials and execution of alleged sorcerers—many of whom were women.

The sixteenth century brought about the beginning of a new era in geography, religion, the arts, and science. The Renaissance in medicine began around 1543 when Andreas Vesalius (1514–1564) published his texts on anatomy. As the medical profession became more regulated, women did not fare well. By the end of the 1600s, the role of women in medicine was giving way to complete male dominance in obstetrical care (the field of medicine that deals with the female reproductive system) that would last for centuries.

Words to Know

Cesarean section: An operation in which a baby is surgically removed from its mother's uterus.

forceps: A medical instrument, shaped like tongs, for extracting a baby during difficult births, as well as for other surgical applications.

gynecology: The branch of medicine concerned with the diseases and physical care of women.

hysterectomy: An operation in which a woman's uterus is removed.

menstruation: The monthly discharge of blood and other materials from the uterus of a nonpregnant female primate (a class of animals including apes as well as humans) of breeding age.

midwife: A person who assists women in childbirth. In premodern times, midwives were almost always female and seldom had formal medical training.

obstetrics: The branch of medicine concerned with childbirth.

ovary: The part of the female anatomy in which eggs (which can be fertilized by sperm to create offspring) are produced.

uterus: An organ in the body of a female mammal, which holds offspring during their stages of development prior to birth.

vagina: A canal leading from the exterior of a woman's body to the uterus.

Early ideas of gynecology and obstetrics

The Greek physician Soranus of Ephesus (flourished in the second century C.E.), wrote the first important Western study on the subject of the female reproductive system: *Gynaecology* (The study of women). Soranus, who concluded that the female body was an imperfect version of the male, was also the first to assert that certain mental disorders were exclusively the property of women. Thus the term *hysteria* is derived from the Greek word for womb: hystera.

Life Science

ESSAYS

Much more constructive was the contribution made by Paul of Aegina (c. 625–c. 690), a Byzantine physician whose great work *Epitomae medicae libri septem* (Medical compendium in seven books) shows him as the leading medical mind of his day. From his discussion of pregnancy and diseases of women, it appears that Paul had experience in a field usually limited to midwives.

Since trained doctors were almost all male, and since few women felt comfortable sharing the intimate details of pregnancy with a man, the job of midwifery (one who helps women in childbirth) fell to women. Paul, however, gained the status of a "man-midwife" and dispensed valuable information regarding subjects such as turning a baby that is positioned incorrectly in the uterus.

Women in the health sciences

Despite all the challenges facing them, some women participated in the medical profession as something other than midwives in premodern times. Not as much is known about these women, however, as is known about the men who practiced medicine. There is some evidence of two women, Agamede and Agnodice, who practiced medicine in ancient Greece, although some scholars argue that accounts of them are fictional.

There is somewhat more information available about Philetas, a medical lecturer in Alexandria, Egypt. At least historians have assigned dates to her life (c. 330–c. 270 B.C.E.). The only other thing known about her, however, is the legend that she was so beautiful she had to speak to her students from behind a curtain. (The same story is also told of the mathematician and philosopher Hypatia [c. 370–415], another female thinker from Alexandria.) Women physicians were numerous during Roman times (c. 200 B.C.E.–c. 500 C.E.). Roman records attest to gynecological work done by both obstetrics, or midwives, and *medicae*, female doctors.

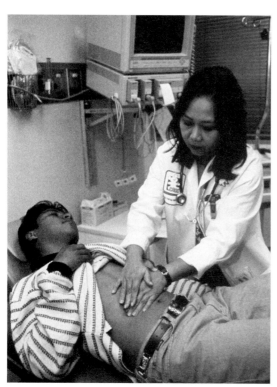

A female physician examines a patient. Women have long practiced medicine, but in premodern times dealt primarily with childbirth and conditions of the female reproductive system. (Reproduced by permission of Photo Researchers, Inc.)

The Middle Ages

Another important early figure—and one whose existence is much more certain—was the Roman noblewoman Fabiola (died c. 399 C.E.), who founded the first civilian public hospital in western Europe in 394 C.E.

(see sidebar on p. 76). For centuries after Fabiola, however, there is little record of female participation in medicine until the 1000s, when the great medical school at Salerno, Italy, emerged (see essay "The Reawakening of the Life Sciences in Europe" in this chapter). Some of the writings studied there were attributed to a woman named Trotula. Some scholars have suggested that Trotula was not responsible for these writings, while others suggest she may also have been a fictional or composite character. It is certain, however, that a number of women studied at Salerno.

Probably the most important woman in the medieval health sciences was Hildegard von Bingen (1098–1179), who practiced herbal medicine. In her book *Physica*, or *The Book of Simple Medicine*, she combined traditional lore concerning herbal treatments and folk medicine with the usual medieval smattering of superstition. In the years that followed Hildegard's time, women were called upon to provide medical treatment during times of great crisis, such as during the Black Death (1347–51), an epidemic also called the plague, but this was primarily because there were not enough male doctors to attend to the population. When the plague passed, women were once again ignored by established medicine, and herbalists eventually came to be regarded as witches.

Yet even during those times, midwives worked behind the scenes, bringing children into the world. They also treated a variety of female illnesses, cared for infants, and prescribed potions to renew sexual desire or encourage fertility. Families of a prospective bride even called upon the midwife to certify the girl's virginity. If the husband-to-be came from a wealthy family, the parents would typically insist that the bride prove her virginity before allowing their son to marry her.

The Renaissance

With the changes in society brought about by the Renaissance (about 1350–1600), women gained an opportunity to receive a medical education. Beatrix Galindo (1473–1535), for instance, even earned certification as a professor at the University of Salamanca in Spain, and the powerful Medici family of Florence, Italy, hired Cassandra Fidelis to translate medical texts for them. Cassandra Fidelis also wrote a book on the natural sciences and treatment of diseases in 1484. In Switzerland, the renowned surgeon Girolamo Fabrici (1537–1619) taught surgery to his wife, Marie Colinet, and later admitted that she was a better surgeon than he was.

Yet even as these positive signs of change began to appear, there was a backlash against women in medicine. Women who practiced herbal medicine, and in some cases even midwifery, were often persecuted as witches. Many innocent women were tortured into confessions of witchcraft, then executed, creating a situation in which women were reluctant to partici-

Life Science

ESSAYS

Life Science

ESSAYS

pate in the medical profession even as midwives. At the same time, university-trained doctors finally began to take an interest in obstetrics and the female reproductive system.

Obstetrics in the 1500s and 1600s

Even as the Renaissance was beginning in about 1350, a shift occurred in the attitudes of the male medical and political establishment of Europe. For many centuries they had more or less left midwives alone, treating the work of the midwife as a necessary task set aside for female practitioners. But as medicine became more formalized and regulated, new laws sprang up governing the role of midwives. During this time, a number of local ordinances limited the function of the midwife, for instance by requiring that she submit to the guidance of a physician or surgeon in difficult births. In 1540 the Guild of Surgeons (a guild was an early form of labor union) declared that "no carpenter, smith, weaver, or woman shall practice surgery."

In the short run these and other laws had a negative effect on the role of women as medical practitioners, but in the long run they signaled changes that would benefit women as medical patients. During the 1500s and 1600s, obstetrics attracted the attention of distinguished physicians such as Ambroise Paré (1510–1590), surgeon to the king of France. By developing a nonsurgical method of turning a baby during a difficult birth, he saved the lives of numerous children and mothers. Later, François Mauriceau (1637–1709) built on Paré's findings and pioneered a technique for delivering babies in the breech, or feet-first, position.

Breech babies and cesarean section

The normal position for a baby in delivery is head-first; when a baby is in the breech position, this poses grave dangers to both the mother and the child. Today these dangers are overcome by techniques such as turning the baby, first suggested by Greek surgeon Paul of Aegina (c. 625–c. 690), and by cesarean section, an operation in which the baby is surgically removed from the mother's uterus. The term "cesarean" refers to the great Roman leader Julius Caesar (100–44 B.C.E.), who was supposedly delivered in this fashion, but the story of Caesar's birth is undoubtedly a legend. Until the early modern era, cesarean sections were only performed to save a living baby after the mother had died in childbirth.

Paré's contemporary François Rousset (1535–1590) was the first to suggest that cesarean sections could be performed on living women. With this idea, he was ahead of his time—perhaps too far ahead of his time. With the still-primitive surgical techniques in use during the 1500s, cesareans were likely to be fatal to the mother. Only in the late nineteenth century, by which time doctors had come to understand the importance of providing an antiseptic (germ-free) environment, did cesarean sections become practical.

Life Science

ESSAYS

Further developments

Gabriele Falloppio (1523–1562), a student of the great anatomist **Andreas Vesalius** (1514–1564; see biography in this chapter), discovered a set of trumpet-like tubes that carry a fertilized egg from the ovary to the uterus. Falloppio himself did not understand the function of these passageways, today known as the fallopian tubes; nonetheless, his description of their structure was highly accurate. He was also the first physician to recognize that the canal leading from the exterior of a woman's body to the uterus is a structure separate from the uterus itself, and he gave that canal its name: *vagina,* from a Latin word meaning "sheath."

The English physician Peter Chamberlen the Elder (1560–1631) wrote on gynecology and obstetrics, but he is best remembered for the invention of forceps. Chamberlen used the forceps, an instrument shaped like tongs, for extracting a baby during difficult births, but in time forceps would gain a wide range of surgical applications. However, the Chamberlen family—who had a hundred-year tradition of delivering babies dur-

Three women are burned at the stake as witches. During the Renaissance women who practiced herbal medicine and midwifery were often accused of being witches. (Reproduced by permission of Archive Photos, Inc.)

Fabiola

A woman of the Roman nobility, or upper class, Fabiola, who died about 399 C.E., is remembered as the founder of the first civilian public hospital in western Europe. Although her family was wealthy, Fabiola, a devout Christian, spent her entire adult life caring for the poor. In addition to the hospital, she cofounded Europe's first hospice, a place where pilgrims and travelers could rest.

Married at first to a violent and abusive man, Fabiola divorced him in accordance with Roman law, and while her first husband was still alive, she married a second man. Fabiola had meanwhile become a Christian and a follower of the early church leader Saint Jerome (c. 347–c. 419). Her second marriage went against the laws of the Christian church. After her second husband died she resolved to atone for what she believed was the sin of remarrying while her first husband was still alive. Therefore, she renounced all worldly pleasures and devoted herself to the care of the poor and the sick.

In 394 Fabiola financed the construction of the hospital in Rome and also offered up her villa (a mansion in the country) as a rest home for discharged hospital patients. In addition, she worked tirelessly, tending to the wounded and diseased on a daily basis.

She later traveled to Bethlehem to study the Christian scriptures in the town of Jesus Christ's (the founder of Christianity) birth, and on returning to Rome she collaborated with a former Roman senator to establish the hospice. In addition to accommodating Christian pilgrims coming to Rome, the hospice, which may have contained as many as four hundred beds, provided for the care of the sick.

Fabiola is reported to have walked the streets of Rome in search of the sick, the dying, and the abandoned, sometimes carrying them to the hospital on her own shoulders. Her life became a model of Christian love and charity, and when she died thousands thronged to her funeral, where Saint Jerome himself preached.

Life Science

ESSAYS

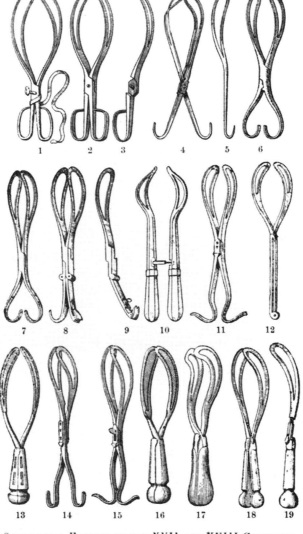

OBSTETRICAL FORCEPS OF THE XVII AND XVIII CENTURIES.
Figs. 1–5. Chamberlen's forceps and vectis ca., 1600; fig. 6. Giffords' extractor, 1726; fig. 7. Chapman's forceps, 1735; figs. 8–9. Freke's, 1734; fig. 10. Mesnard's, 1741; fig. 11. Grégoire's, 1746; fig. 12. Rathlauw's, about 1747; fig. 13. Pugh's, 1754; figs. 14–15. Levret's, 1747; fig. 16. Smellie's, 1752; fig. 17. Johnsen's, 1769; fig. 18. Fried's, 1770.

Obstetrical forceps of the seventeenth and eighteenth centuries. Invented by English physician Peter Chamberlen the Elder for extracting a baby during difficult births, forceps eventually had many surgical applications. (Reproduced by permission of Hulton/Archive.)

ing both normal and difficult births—closely guarded their invention. Even the mothers on whom the forceps was used never saw the instrument, nor did the Chamberlen family permit midwives to use them.

Thanks to the Dutch anatomist Reinier de Graaf (1641–1673), who gave the ovary its name, gynecology progressed from simple observation

Life Science

ESSAYS

of structures, as in the case of Falloppio, to an understanding of those structures' functions. Compared to the writings and the works of his contemporaries, de Graaf's work was much more advanced. While the drawings of others were inexact and full of speculation, de Graaf—who illustrated his own texts—offered precise and detailed pictures.

The famous midwives

At the same time that these advances in obstetrics were occurring, a strange thing happened. Midwives had previously been anonymous, behind-the-scenes figures, at the fringes of the medical mainstream. They had been pushed even further to the side with the growing medical interest in childbirth, as the tide of law turned against their practice. Yet the period from the 1500s to the 1700s saw the rise of several women who not only practiced midwifery, but wrote about it and became famous for their work.

Among these was Louise Bourgeois (1563–1636) of France, midwife to the Medici family (a family that was dominant in Italian politics for two and a half centuries), as well as England's Jane Sharp (flourished about 1650). Both sought to provide midwives with sound information on anatomy and the techniques of midwifery, but the English-French midwife Elizabeth Cellier went even further. In 1688 she proposed the idea of a college to provide instruction and supervision to midwives.

Obstetrics in the 1700s

The authorities never approved of Cellier's idea, however, and events increasingly pointed toward the end of midwifery as a practice. As Chamberlen's descendants released their exclusive claim over the forceps, and use of the instrument spread to other medical professionals, this greatly diminished the dominance of midwives in obstetrics. By 1750 a number of physicians and surgeons had gained the status of "man-midwives," and the growth of university courses on obstetrics established it as a distinct medical specialty.

At the forefront of this movement was Scottish physician William Smellie (1697–1763), who learned obstetrics by providing free medical care to poor women in London, England. Smellie, who greatly improved the design of the forceps, published one of the most significant works in the history of obstetrics, *Treatise on the Theory and Practice of Midwifery*, in 1752.

By the late 1700s, most upper-class women relied on professionally trained doctors rather than midwives; yet in America, where doctors were more scarce than in Europe, the profession of midwife continued to flourish into the 1800s. Furthermore, as a result of rising interest in home childbirth and alternative medicine during the late twentieth century, midwives have again become common throughout the Western world.

☐ BIOLOGY AND CLASSIFICATION

Life Science

ESSAYS

Progress in the field of biology, with its related areas of study in botany and zoology, has had an enormous indirect impact on everyday life. Indeed, the practice of medicine in premodern times relied heavily on studies in these areas. For instance, most of the medications prescribed during the Middle Ages (roughly 500 to 1500 C.E.) were herbal, or plant-based, treatments whose development benefited from the work of ancient botanists who studied plants and their effects on humans. In addition, the system for classifying species established by Swedish botanist **Carolus Linnaeus** (1707–1778; see biography in this chapter), greatly enhanced scientists' understanding of human beings and their place in nature. This in turn enabled the advances of the 1800s that would lead to the birth of the modern life sciences.

Aristotle and the origins of biology
Aristotle's (384–322 B.C.E.; see biography in this chapter) detailed study of the natural world earned him the title "father of biology," yet the impact of

"The School of Athens" mural, painted by Santi Raphael, depicts Plato and his student Aristotle who together developed the methods of reasoning that formed the foundation of science, the arts, and politics in the West. (Reproduced by permission of Erich Lessing/Art Resource, New York.)

Words to Know

bestiary: An ancient or medieval catalogue of animal life. Bestiaries were highly unscientific and typically included numerous fictional creatures such as the unicorn.

biology: The scientific study of living organisms. Actually a collection of disciplines that includes botany and zoology, biology is (along with medicine) one of the two principal areas of study in the life sciences.

botany: A branch of biology concerned with plant life.

genus: A group of species distinguished by common characteristics.

herbal: An ancient or medieval book concerned with plants and their medicinal uses.

species: A category of closely related organisms. A species is usually defined by the ability of its members to breed with one another and by their inability to breed with members of other species.

zoology: A branch of biology concerned with animal life.

this Greek philosopher on Western thought has been so enormous that his contributions to biological study are far from the greatest of his accomplishments. Indeed, without Aristotle and his teacher Plato (c. 428–348 B.C.E.), it is hard to imagine how the "West" as such would have emerged. Together they developed methods of reasoning that formed the foundation of the sciences, the arts, and politics.

A central aspect of Aristotle's philosophy was the idea of classification—providing an organizational structure to a variety of ideas. He was more than just a thinker, however; at the school he operated in Athens, Greece, he and his students conducted practical research in the varieties of animal life, examining some five hundred species. Nor did they simply record data: the purpose of their work was to develop an understanding of each animal's relationship to the larger environment. This, of course, is a very modern idea—like Aristotle, today's environmentalists maintain that everything in nature is somehow tied to everything else.

An illustration from a fourteenth-century herbal. Herbal medicine was the primary method of health care in the Middle Ages, and herbals were the most significant medical textbooks of the time. (© Gianni Dagli/Corbis. Reproduced by permission of the Corbis Corporation.)

Life Science

ESSAYS

Theophrastus and other early botanists

Just as Aristotle established the science of biology, his student Theophrastus (c. 372–c. 287 B.C.E.) is known as the "father of botany." In *Inquiry into Plants,* he described not only plants native to Greece, but also species from around the Mediterranean, the Atlantic Coast, and India.

Theophrastus's *De Causis Plantarum* presented observations on plant reproduction, including an accurate description of the germination of seeds. Using an Aristotelian approach, he emphasized direct observation, and instead of forming broad generalizations from little evidence—a pitfall

Life Science

ESSAYS

of many scientists in the Middle Ages—he only made general statements when he believed he had enough specific facts to support them.

No botanist of Theophrastus's caliber would arise for the next eighteen hundred years; however, a few scientists of note did follow him in the first century. The Roman Pliny the Elder (23–79) wrote *Historia Naturalis* (Natural history), which, though not as well organized as Theophrastus's work, would have an enormous impact on medieval botany. Pliny's work was the only satisfactory presentation of the botanical ideas of Theophrastus to survive into medieval times. Sadly, the writings of Theophrastus were lost and only rediscovered in the 1400s. Also notable was Pedanius Dioscorides (c. 40–c. 90), whose writings were widely studied by medieval botanists. A few notable female scientists of the medieval period made contributions to botanical study, including Hildegard von Bingen (1098–1179) who produced over two hundred works, including the *Physica*, an herbal that discussed plants and their medicinal uses.

Herbal medicine in the Middle Ages

During the Middle Ages virtually the only forms of medicine were herbal in nature, and the most significant medical textbooks of the day were herbals. Like Hildegarde's *Physica*, these contained descriptions of plants and their medicinal uses, but much of the information presented in medieval herbals was inaccurate.

Medieval biology comes of age

The work of French philosopher Peter Abelard (1079–c. 1144) on plants is notable for its improved scientific reasoning that resulted from exposure to the ideas of Muslim scientists (see essay "The Great Doctors of Medieval Islam" in this chapter). Likewise, the writings of Albertus Magnus or Albert the Great (c. 1200–1280) hinted at the increased understanding of botany then filtering from Muslim scientists into the world of European scholars.

In *De vegetabilibus de plantis*, Albertus provided highly accurate and detailed descriptions of the medical and economic uses of some 270 plants. Like Abelard and virtually all medieval scholars, Albertus was a priest, and in his work he traveled throughout Germany on foot. This gave him opportunities to observe plants in their natural environments, and his writings discussed the ways that species found in forests differ from those found in swampy areas and open fields.

Albertus's work in zoology

However, Albertus's contributions were not limited to botany: in his vast work *De animalibus* he provided extensive discussions of animal reproduction. Though he primarily summarized ideas from past writers such as

Botanical and Zoological Gardens

Almost everyone has, at one time or another, visited a zoo or a botanical garden. Few people, however, realize that these have their roots in studies of plant and animal life that took place during the Renaissance (the period of artistic and intellectual rebirth, lasting roughly from 1350 to 1600, that marked the end of the Middle Ages). Even prior to that time, however, kings and nobility kept collections of plants and animals for much the same purpose that people visit zoos and botanical gardens today—for educational entertainment.

Menageries, an early form of zoological garden (the term has long since been shortened to "zoo"), were simply collections of animals, not organized according to any scientific principle. For instance, Henry III (1207–1272) of England kept a menagerie, which included camels, lions, leopards, and the first elephant ever seen in the country. The only medieval menagerie with a scientific purpose was that of Holy Roman Emperor Frederick II (1194–1250), an enthusiastic supporter of the sciences who even wrote a study on the sport of falconry, the art of hunting with birds of prey.

The discovery of new animal species in the Americas heightened interest in botanical gardens during the late Renaissance. The first botanical garden was founded in Padua, Italy, in 1545, and by the end of the 1600s botanical gardens had sprung up in cities all over Europe. The fact that many of the plants were tropical in origin and not compatible with the European climate led to the development of heated greenhouses.

European scholars were interested in plants because they offered potential applications in medicine and other areas; by contrast, the many animal species discovered in the New World held out no such promise. Therefore, until the nineteenth century, there were few actual zoological gardens (as opposed to menageries) in Europe. An exception, however, was the royal zoological garden of Sweden. Established in 1561, it had a great influence on **Carolus Linnaeus** (1707–1778; see biography in this chapter) in developing his system for classifying animal and plant life.

Life Science

ESSAYS

Aristotle, the book is nonetheless considered a turning point in medieval zoology because it greatly exceeded the writings that preceded it.

Prior to Albertus's time, the literature of medieval zoology had been confined to *bestiaries,* entertaining if highly inaccurate catalogues of fanciful creatures such as the unicorn. Albertus, on the other hand, called for first-hand observation—not imagination or the tall tales of those who had journeyed to foreign lands—as the basis for discussions of animal life.

Biology in the Renaissance

A number of other figures continued to further the emerging science of zoology in the centuries that followed. With the opening of international trade routes and the exploration of the Americas, a great deal of additional information about exotic species came to light. In the 1500s Swiss scientist Conrad Gesner (1516–1565) made another milestone in the history of zoology with his illustrated five-volume work *Historiae animalium.* The work included descriptions of many animals never before seen by most Europeans and denounced the practice of including fictitious animals in bestiaries.

Meanwhile, the invention of the printing press in about 1450 made possible for the first time the production and widespread distribution of illustrated books. (See essay "The Printed Word" in the Technology and Invention chapter.) The first important work produced in this period was *Herbarum vivae eicones* (Living images of plants) by Otto Brunfels (c. 1488–1534), soon followed by Leonhard Fuchs's (1501–1566) *De historia stirpium.* These are considered among the most finely illustrated botanical books ever produced, and they set the standard for botanical illustration. (See essay "Printing, Illustration, and Dissection" in this chapter.)

The discoveries of new plant and animal species in the Americas raised an issue that had not been seriously approached since the time of Aristotle and Theophrastus: the need for a consistent method of classifying, organizing, and naming species. The first scholar of the modern era to attack this problem was the Italian botanist Andrea Cesalpino (1519–1603), but nearly two centuries would pass before the development of a workable classification system. Success in this endeavor came in the 1700s, with the work of Carolus Linnaeus. At this time scholars still wrote mainly in Latin, not because they were trying to adhere to tradition but because it remained a common language between educated people of different countries. Thus Linnaeus's great work of 1735 was named *Systema naturae* (The natural system).

The language of ancient Rome also played a key role in Linnaeus's classification system, which gave two Latin names to each variety of plant or animal. The first name, capitalized and sometimes abbreviated, identifies it by its generic name, its genus, or family; the second name identifies it specifically, its species. This method, known as binomial nomenclature, remains in use today, as do the Latin names themselves—even though English, and to a lesser extent French, has replaced Latin as the language of international scientific communication.

How the system is organized

The generic name made it possible for botanists and zoologists to identify closely related life forms, while the specific name identified particular organisms. But genus and species are only two levels in Linnaeus's vast system, which first divides all living things according to kingdom—for instance, plant or animal. Below the level of kingdom are those known as phylum, subphylum, class, order, family, and finally genus and species.

Humans, or *Homo sapiens*, belong to the animal kingdom (Animalia); the Chordata (that is, possessing some form of central nervous system) phylum; and the Vertebrata subphylum, indicating the existence of a backbone. Within the mammal (Mammalia) class, humans are part of the primate (Primata) order, along with apes. Humans are further distinguished as members of the hominid (Hominidae) family, the genus *Homo* ("man"), and the species *sapiens* ("wise").

The impact of Linnaeus's system

The classification system described here is not purely the work of Linnaeus; others, including French naturalists Baron Georges Cuvier (1769–1832), Michel Adanson (1727–1806), and Georges-Louis Leclerc de Buffon (1707–1788) refined it greatly. Nevertheless, Linnaeus is usually given credit for the overall system, officially approved by the international community of zoologists and botanists in the early twentieth century.

The use of a single, universal scientific name for each species constituted a giant step forward for zoologists and botanists. From then on scientists around the world used the same name for the same species and could share data, confident that they were working with the same organism. Furthermore, the Linnaeus system made it easier for scientists to express hypothesized relationships between various types of organisms, which was critical to the evolutionary theory of Charles Darwin (1809–1882). Thanks to Linnaeus, the biological sciences had achieved what Aristotle sought two thousand years before: a uniform system of classification.

Life Science

ESSAYS

☐ THREE NEW WORLDS

In the mid-1400s sailors from Europe began embarking for lands previously unseen by Europeans—first in Africa, and later in the Americas, or

Life Science

ESSAYS

the New World. When Europeans began exploring the American continent's interior in the 1500s, they saw the vast land as a source of new plants, animals, and minerals to use and to transport back to Europe. As they colonized this New World, they also brought with them many familiar plants and animals for food, farming, and other purposes. This exchange of species between the two continents had positive and negative effects. On the positive side, the exchange introduced what would become important agricultural crops and beneficial animals to both continents. It also, however, expanded the range of species that carried disease and competed with beneficial native species and permanently changed the face of each continent. Contact between Europeans and Native Americans led to a demographic disaster of unprecedented proportions. Many of the epidemic diseases that were well established in Europe were absent from the Americas before the arrival of Christopher Columbus in 1492. Influenza, smallpox, measles, and typhus fever were among the first European diseases imported to the Americas—diseases that would decimate the Native American population.

The great geographical space of the Americas was only one of several new worlds explored by Europeans from about 1450 to 1800. Even as the colonization of the Western Hemisphere introduced new forms of plant and animal life, scientists discovered a much smaller, even more complex and varied realm, with the aid of the recently invented microscope. Coinciding with these discoveries was a third form of exploration, a journey into the most intimidating terrain of all—the human mind.

The Americas

When Christopher Columbus (1451–1506) landed in the New World in 1492, he discovered an untamed continent brimming with mysterious plants and animals. Much has been written about the misfortunes the Europeans brought with them to the New World: the destruction of the Aztec and Inca empires in Mexico and South America, as well as other native populations; the conquest of the Indians' land and the failed attempt at enslaving them; and finally the replacement of Indian slaves with those captured from Africa. Less well-known, however, is the impact of European exploration on the sciences, much of which was positive—even, in some cases, for the American Indians themselves.

Some early European settlers showed the Native Americans how to increase their crop yields, and the Native Americans in turn introduced the colonists to numerous vegetables the colonists had never seen before. Among these were corn (or maize, as the Indians called it), as well as the potato, first cultivated in South America. Sailors brought samples back home with them, and soon crops of corn and potatoes sprang up around Europe.

Life Science

ESSAYS

Spanish conquistadors clash with Aztec soldiers. Much has been written about the misfortunes Europeans brought with them to the New World. Less well-known, however, is the impact of European exploration on the sciences, much of which was positive. (Reproduced courtesy of the Library of Congress.)

Because Italian cooking today is so noted for its many tomato-based sauces, it may be hard to imagine a time when the tomato was new to Italy, but it caused a sensation there when sailors brought it back from Central America in the 1500s. Many Italians, believing that the strange new fruit possessed the power to stimulate sexual desire, dubbed it the "love-apple." In addition to such edible plants, the Indians introduced explorers to tobacco, which became extremely popular in Europe—though even then, many people disapproved of its use.

What Europeans brought to the Americas

Where animal life was concerned, most of contributions made by Europeans were welcome, such as introducing Native Americans to cattle, pigs, chickens, goats, and sheep. Of equal importance was the horse. Until the arrival of the Europeans, the only domesticated work animal in the entire New World was the llama, a small South American cousin of the camel. The Inca had used llamas to carry loads over the high Andes Mountains of South America, but the creatures were not large enough to provide human transportation.

Europeans also brought to the New World such Old World crops as wheat; however, they unintentionally carried with them a number of harmful species. Among these were rats and other pests that made their way onto ships bound for the Americas, as well as dandelions and other weeds, which soon took root in the new colonies. None of these visible species, however, caused nearly as much damage as the invisible ones living in the bodies of the Europeans themselves.

Life Science

ESSAYS

The spread of diseases

Due to the centuries of interaction between populations in Europe, North Africa, the Middle East, and the Far East, the peoples of the Old World had developed a high level of immunities to various disease-causing germs. By contrast, the natives of the New World—though they had their own share of illnesses—never had been exposed to a wide range of diseases that included smallpox, measles, chicken pox, influenza, typhoid fever, and the bubonic plague.

When the bubonic plague and related forms of disease had struck Europe in the Black Death (1347–51), it killed as much as 30 percent of the population. From 1492 to 1650 perhaps as much as 90 percent of the native population in the New World succumbed to epidemics, as well as to food shortages caused by the Europeans' conquest of their lands.

The Microscope

The Europeans who came to settle the New World had no idea they were carrying a powerful form of biological warfare—communicable diseases. At the time, even the most advanced scientists had only a vague idea of the causes behind diseases: no one, after all, could see bacteria. Only in the nineteenth century would scientists develop germ theory, an explanation of disease and infection as the result of exposure to harmful microscopic life forms. When Europeans first landed in the New World, the microscope had yet to be invented.

Scientists in ancient Assyria (now part of Iraq) and Greece had experimented with the use of lenses for magnification, but only with the invention of eyeglasses in about 1300 did research in the use of magnifying lenses make significant progress. Near the end of the 1500s, it became apparent that if certain lenses were joined together by a cylinder, they could be used either as a telescope or a microscope, depending on which end was used to view objects. The great astronomer **Galileo Galilei** (1564–1642; see biography in Physical Science chapter), was one of the first to use the telescope to see very large, faraway objects, but only in the 1600s did scientists use lenses to view very small ones.

Four early microscopists

Credit for the early development of the microscope as an instrument for use in the life sciences goes to four men: Italian physician and botanist Marcello Malpighi (1628–1694); Dutch biologist Antoni van Leeuwenhoek (1632–1723); Dutch anatomist Jan Swammerdam (1637–1680); and English physicist and physician Robert Hooke (1635–1703).

Malpighi was one of the first scientists to use a microscope to study the structure, composition, and function of living tissue; and Leeuwenhoek became the first to view single-celled organisms. His countryman Swammerdam used the instrument to observe a variety of creatures, from spiders to snails, while Hooke first identified and named cells, which he discovered in a piece of cork.

In his highly influential *Micrographia* (1665), Hooke described in detail the structure of feathers, the stinger of a bee, and the legs of a fly. He also noted that certain plant cells seemed to be filled with liquid and suggested their function was to transport substances throughout the plant. His work, as well as that of the others, paved the way for enormous advances in the life sciences during the centuries that followed.

Exploring the Human Mind

If scientists' understanding of microscopic structures was elementary, their knowledge in another area—that of the human mind—was downright primitive. This, of course, is highly ironic, given the fact that it was the mind that enabled them to make all preceding discoveries. Yet even as settlers were conquering the New World, the human mind remained a great unexplored frontier.

Actually, the mind (as opposed to the brain) is somewhat beyond the reach of biology and medicine. These disciplines, along with physics, chemistry, and other "hard sciences," are devoted to things that can be measured and predicted. By contrast, the mind—as the seat of creativity, the will, the emotions, and other characteristics that make each human individual—remains largely a mystery.

The human brain is a physical structure whose location in the head is obvious; on the other hand, no one really knows where (or what) the mind is. Indeed, it might even be more properly called the soul, a term that is clearly outside the realm of science. Thus religious teachers, writers, and artists are at least as qualified as scientists to discuss the workings of the mind (not to be confused with the brain).

If the mind can be understood at all from the standpoint of science, it must be approached by those in the social sciences, most notably psychology. Modern psychology incorporates elements of biology and medicine, but it is also concerned with concepts, such as the effect of childhood experiences on a person's behavior as an adult, that are difficult to measure.

Life Science

ESSAYS

Robert Hooke's microscope. Hooke first identified and named cells in 1665. ©Bettmann/Corbis. Reproduced by permission of the Corbis Corporation.)

Life Science

ESSAYS

By contrast, the modern discipline of psychiatry is a branch of medicine. A psychiatrist is legally qualified to prescribe drugs—in this case, treatments that directly alter the chemistry of the physical brain. Another branch of medical science related to the study of the brain is neurology, which is concerned with the nervous system.

Diagnosing and treating mental illness

The greatest Greek philosophers each had highly sophisticated theories about the way that the mind interprets the world (see sidebar), but Greek scientists possessed virtually no understanding of the brain. **Hippocrates** (c. 460–c. 377 B.C.E.; see biography in this chapter) explained mental illness in terms of the four humors, the belief that human health is governed by a balance of four bodily fluids.

There were two types of mental illness, Hippocrates wrote: mania or insanity, which resulted from an excess of choler or yellow bile; and melancholy or depression, caused by too much black bile. To cure these problems, he prescribed bleeding and purging (forced vomiting), as well as rest and exercise.

Mentally ill persons with violent tendencies, of course, posed a threat to the community, and both Greek and Roman societies made provisions for separating these individuals. There were no mental hospitals, however, and therefore these patients were the responsibility of their families.

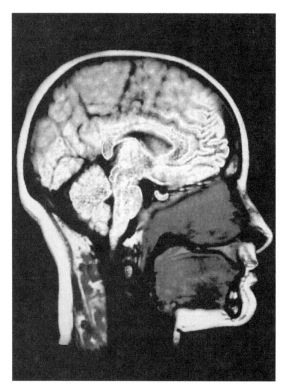

MRI scan of the human brain. The location of the human brain is obvious, but the mind as the seat of creativity, the will, emotions, and other characteristics that make each human individual remains largely a mystery. (Reproduced by permission of Photo Researchers, Inc.)

The Middle Ages

The analysis and treatment of mental illness did not improve during the Middle Ages (roughly 500 C.E. to 1500 C.E.); if anything, it got much worse. The *best* forms of medical care available were based on humoral theory, and superstitions abounded. Thus arose the popular belief that the Moon (*luna* in Latin) caused insanity, or lunacy. There was also the notion, promoted by the church, that insanity was a sign of demonic possession.

Not all unscientific beliefs in that era were negative, however. Some communities viewed those who were mentally ill or mentally retarded (medieval people had no concept of the difference) as special children of God. Many even believed that persons with mental illnesses possessed

Life Science
ESSAYS

what would now be called psychic abilities, or special gifts of insight. This led to the idea of the "wise fool," a popular literary theme ever since—reflected, for instance, in the 1994 Academy-Award-winning motion picture *Forrest Gump*.

For the most part, however, people treated the mentally ill and the mentally retarded with hostility. Some communities whipped them or ordered them to leave, while others established "madman towers" to lock them away from society. While madman towers were barbaric, this practice led to the establishment of the first psychiatric wards, as well as hospitals for the insane. These hospitals, however, were far less than adequate. An example was Saint Mary of Bethlehem, established in London, England, in 1247, where conditions were so miserable that a slurred version of the hospital's name, "bedlam," entered the language as a synonym for confusion.

In 1484 the pope, leader of the Catholic Church, announced a campaign to root out witches throughout Europe. In the years that followed, numerous persons with mental problems—as well as a host of others who exhibited unusual behavior or ideas—would die by torture or execution. The status of the mentally ill did not improve until the 1700s, thanks to a series of scientific discoveries that pointed to physical, rather than spiritual, explanations for the brain's functions.

Neurology and the functions of the brain

These changes had their roots in the new science of neurology, a term coined by English physician and anatomist Thomas Willis (1621–1675). Willis belonged to an emerging school of thought known as *iatrochemistry*, whose members maintained that chemical interactions governed the functions of human and animal bodies. He reasoned that something in the brain causes the body to move, but it was his countryman Stephen Hales (1677–1761), a minister, who conducted the first scientific studies of nerve function.

Whereas most physicians of the day believed that blood pressure caused muscle contractions, which are necessary for movement, Hales suggested that electrical impulses are actually responsible. Experiments with frogs led him to the conclusion that animals possess a nervous system linked to the spinal cord. Italian physician Luigi Galvani (1737–1798) continued this work and proposed a theory of animal electricity, the idea that an electrical current of some sort runs through an animal's body.

Later the Italian physicist Alessandro Volta (1745–1827), for whom the unit of electrical measure known as the volt is named, found that excessive electrical stimulation can cause a muscle to contract continually. This is indeed what happens when a person is exposed to a strong electric current, and it explains why such exposure can be fatal.

Life Science

ESSAYS

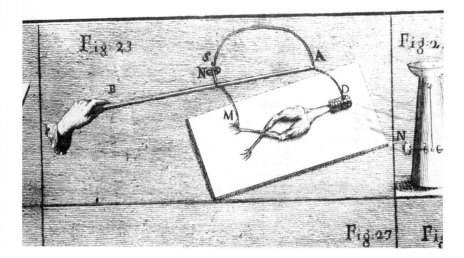

An engraving of Luigi Galvani's experiment with frog legs. Galvani's experiments led him to the conclusion that animals possess a nervous system linked to the spinal cord. (UPI/Corbis-Bettmann. Reproduced by permission of the Corbis Corporation.)

Effects of animal electricity theory

Though the idea of a special animal variety of electricity has long since been discarded, scientists today recognize the fact that electricity does run through the human body, sending signals from the brain to the nerves and from the nerves back to the brain. Thus in the long run, the work of Galvani and others led to enormous progress in the field of neurology.

The short-term effect, however, was to inspire a number of fake miracle cures, whereby static electricity was used to treat illnesses. This practice later influenced the use of shock therapy for mental patients. Contemporary ideas concerning the relationship between electricity and the functions of the body are also reflected in the 1818 novel *Frankenstein* by Mary Shelley (1797–1851). The tale, familiar to anyone who has seen one of the many movies based on it, concerns a doctor who uses electricity to create life.

☐ THE HUMAN MACHINE

Throughout much of history, scientists have not simply faced the challenge of understanding how nature works; they have also had to respond to the weight of prevailing beliefs, beliefs that have made people reluctant to accept the findings of science. An obvious example is the debate over Charles Darwin's (1809–1882) theory of evolution, with its implication that humans are very intelligent cousins to the apes. Though Darwin based his findings on science, and had no aim of attacking religious beliefs about creation, many critics maintained that evolution makes no allowance for the idea of the human soul or spirit.

Words to Know

embryo: An unborn animal in the early stages of development. For humans, the embryonic stage is defined as the period of eight weeks following conception, or the fertilization of the egg.

pulmonary circulation: The movement of blood from the right ventricle of the heart into the lungs via the pulmonary artery, and from the lungs to the left ventricle through the pulmonary veins.

ventricle: A chamber in the heart.

Throughout the Middle Ages (roughly 500 C.E. to 1500 C.E.) and the Renaissance (the period of artistic and intellectual rebirth, lasting from about 1350 C.E. to 1600 C.E., that marked the end of the Middle Ages) to the dawn of the modern era, there has been tension between the church and the scientific community. The church maintained that the body was a physical reflection of spiritual realities, whereas by the 1500s many scientists had come to see it as a type of machine that could be analyzed and understood like any other mechanism. Even as discoveries regarding human circulation, as well as advancements in surgical techniques, took away some of the mysteries of the human body, this debate raged throughout the 1600s and 1700s.

Iatromechanism versus vitalism

The idea that the body is actually just a complex machine originated in the 1500s and 1600s and was called iatromechanism or iatrophysics. One of its leading figures was French philosopher **René Descartes** (1596–1650; see biography in Math chapter). On the other side of the debate was a movement known as vitalism, whose adherents held that a vital force (that is, a life force) differentiates living things from nonliving ones.

For centuries people have been perplexed by the fact that the physical human body seems to contain an extremely important nonphysical component, variously called the mind, soul, or spirit. "This being of mine, whatever it really is," wrote the Roman emperor and philosopher Marcus Aurelius (121–180) in his *Meditations*, "consists of a little flesh, a little breath, and the part which governs."

Life Science

ESSAYS

The "flesh" part, or the body itself, has long been considered the domain of science: but what about the "breath"—the thing that gives life to an organism such as the human being and "the part which governs"—the will, the intellect, and the emotions? How do the three components interact?

The "rational soul" and the "ferments"

Descartes focused on "the part which governs," and maintained that what he called "the rational soul," or the seat of wisdom, interacted with the body in the brain's pineal gland. According to Descartes, the nerves send signals from the body to the brain, and the rational soul makes split-second decisions regarding its response to those signals.

Vitalists such as Flemish physician and alchemist Jan Baptista van Helmont (1579–1644), on the other hand, were more concerned with the "breath" that separates living creatures from nonliving ones. Only living organisms, van Helmont maintained, could produce the "ferments" that permit the digesting of foodstuffs and their incorporation into the organism.

Evaluating these views

In the nineteenth century, scientists identified van Helmont's "ferments" as enzymes—chemical substances rather than living creatures. As for the idea that there is a sharp dividing line between living and nonliving things, this notion suffered a fatal blow with a demonstration that resulted in the redefinition of the terms "organic" and "inorganic."

Prior to that time, "organic" had referred to anything that was living, formerly living, or part of a living or formerly living thing. Everything else fell under the heading of "inorganic," and the wall that separated the two seemed to be an unbridgeable gulf. In 1828, however, German chemist Friedrich Wöhler (1800–1882) made an astounding discovery while trying to synthesize (manufacture) ammonium cyanate. He treated silver cyanate with ammonium chloride, but instead of ammonium cyanate he ended up with urea, a simple compound found in animal urine. This may not sound amazing, but to Wöhler, it was as though he had created life, because urea comes from living things, whereas living things do not contain silver cyanate or ammonium chloride. As Wöhler eventually realized, all living things and

French philosopher René Descartes was a leading proponent of iatromechanism or iatrophysics—the idea that the body is actually just a complex machine. (Reproduced by permission of the Library of Congress.)

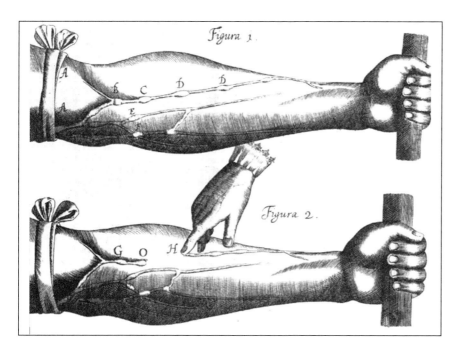

Life Science

ESSAYS

An illustration from William Harvey's "De Motu Cordis" showing that a connection must exist between the arteries and veins. Harvey developed the first accurate description of the human circulatory system.

their products, such as urea, contain carbon and hydrogen chemically bonded. These two elements had joined when carbon from the silver cyanate had bonded with hydrogen from the ammonium chloride. As a result, organic chemistry became identified as the branch of chemistry that deals with compounds (substances that contain chemically bonded elements) that contain carbon and hydrogen. Most of these come from living things, but not all organic compounds do; however, as long as they contain carbon and hydrogen, they are organic.

Advances in the medical sciences

Though Descartes failed to locate the soul, he and the iatromechanists were on the right track when they described the human body as a machine or mechanism, and discoveries by English physician **William Harvey** (1578–1657; see biography in this chapter) reinforced this position. Harvey developed the first accurate description of the human circulatory system, based on his many years of experiments and observations as a scientist and physician.

Likewise, advances in surgery during the 1500s and 1600s seemed to point to the idea that the body could be understood and "fixed" just as one analyzes the workings of, and performs repairs on, a clock. Of the many physicians who developed new surgical techniques during this period, none was more important than Harvey, one of the most important figures

Life Science

ESSAYS

in the history of the life sciences. Yet each contributed to the betterment of human life, as well as to the emerging model of the body as a machine.

Harvey and the circulatory system

Until Harvey's time, beliefs about the workings of the circulatory system had come chiefly from the writings of **Galen** (129–c. 199; see biography in this chapter)—who, incidentally, was Roman emperor Marcus Aurelius's personal physician. Though advanced for their time, Galen's ideas contained many errors: for instance, he maintained that the liver served as important a function in the circulatory system as the heart, and that blood seeped from the right *ventricle,* or chamber, of the heart to the left ventricle through invisible pores. Due to the almost total lack of scientific progress in Europe during the millennium after his death, Galen's work acquired the status of scripture, and physicians challenged it at their own peril.

As the personal physician to the king of England, Harvey was the leading physician of his country, and perhaps in all of Europe. He came to doubt Galen, and in 1628 Harvey published his findings in *Exercitatio anatomica de motu cordis sanguinis in animalius* (On the movement of the heart and blood in animals). The heart, he explained, is the principal organ in the circulatory system. As for the problem of how blood moves from the right ventricle to the left, he established the concept of pulmonary circulation: the movement of blood from the right ventricle into the lungs via the pulmonary artery, and then from the lungs to the left ventricle through the pulmonary veins.

Harvey's methods were as important as his findings themselves. Instead of using the writings of Galen or other physicians of the past as his guide—a tendency that had characterized European medicine throughout the Middle Ages—he relied purely on experimentation and the observations derived from it. Galen's followers did not surrender quietly to Harvey's ideas, but rather they challenged him on every detail of his new theories. As was his tendency, Harvey patiently dealt with these objections, sometimes sending his critics letters in which he explained how he had arrived at his findings. Only after his death did scientists come to accept the truth of his circulatory model.

Advances in surgery

The many wars that took place during the Renaissance (about 1350 C.E. to 1600 C.E.) gave surgeons extensive experience and led to improvements in treating wounds. This resulted in many surgical "firsts" during the mid- to late 1500s: the first successful tracheotomy, or the removal of a blockage in the windpipe; the first reported splenectomy, an operation on the spleen; the first eye removal for cancer; and advancements in plastic surgery.

For example, using techniques first applied centuries before by Indian physicians (see essay "'Alternative' Medicine Begins in the East" in this chapter), Italian surgeon Gaspare Tagliacozzi (1545–1599) removed skin from a patient and fashioned a new nose. He also performed surgery to correct harelips. Tagliacozzi's work resulted in harsh criticism from the church, which banned him from surgery for altering "God's handicraft."

By the 1600s the first successful stomach surgery (the removal of a knife from the stomach of a peasant sword-swallower) had taken place; the first reported surgery for cancer of the tongue and the first hemiglossectomy (partial tongue removal) had been performed; and a number of other advances, including treatments for the infection known as gangrene, had taken place.

Life Science

ESSAYS

Natural theology versus the mechanists

These advances in surgery served to further heighten the debate between those who believed that the body was a mechanism whose workings could be understood, and those who maintained that the body would always remain an object of mystery. In England there arose a movement known as natural theology, which attempted to find proof of God's existence in nature. Natural theology attracted a number of followers among the greatest men of science, including English chemist and physicist **Robert Boyle** (1627–1691; see biography in the Physical Science chapter).

In contrast to the natural theology movement was that of the mechanists, or materialists. Influenced by the work of Descartes, these scientists—among them French physician and philosopher Julien Offfroy de La Mettrie (1709–1751)—believed that all life processes could be explained purely in terms of physical and chemical laws. In *L'Homme-machine* (Man the machine) in 1747, La Mettrie presented man as a machine whose actions were entirely the result of material factors. By the late 1700s, science had largely discarded mechanism or materialism as an oversimplification.

Preformation versus Epigenesis

During the 1600s and 1700s the nature of the human *embryo* was the focus of another great debate that likewise involved some of the most celebrated scientists and philosophers of that time. Many were convinced that God had formed all embryos at the time of Earth's creation and that these simply emerged and developed as fully formed organisms when their time came. Others, however, maintained that each embryo was formed gradually, structure by structure, through natural processes.

The first of these ideas was known as *preformationism,* and its adherents included such distinguished figures as Dutch microscopist Jan

Life Science

ESSAYS

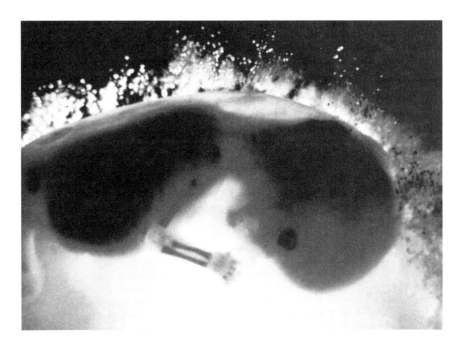

A human fetus at ten weeks. During the 1600s and 1700s the nature of the human embryo was the focus of a great debate that involved some of the most celebrated scientists and philosophers of the time. (© CNRI/Phototake. Reproduced by permission.)

Swammerdam (1637–1680). Preformationists believed that all living organisms, plant and animal, were created in complete but miniature form within the eggs that each parent contained. Their theory of *emboîtement* (embodiment) indicated that the embryo contained all future generations in a long series of miniature, complete embryos that were held in place, awaiting their appointed time to grow and emerge.

Such ideas had circulated from ancient times but were formalized only in the 1700s. When Dutch microscopist **Antoni van Leeuwenhoek** (1632–1723; see biography in this chapter) discovered the existence of spermatic animalcules, or sperm, in 1677, preformationists argued that the sperm held the preformed embryo. Some even imagined that a homunculus, or a tiny human, could be found in the head of the sperm.

The opposing view, called *epigenesis,* also had ancient roots—in this case, with Greek philosopher **Aristotle** (384–322 B.C.E.; see biography in this chapter). Aristotle had observed that the chicken embryo and its internal structures were not preformed but rather formed gradually, and two thousand years later, Harvey (who coined the term "epigenesis") reached the same conclusion. Harvey agreed with Aristotle and maintained that some sort of "life force" initiated the embryo's growth after fertilization. By the 1700s, epigenesis had largely prevailed over preformationism.

DNA raises new questions

Clearly much of the preformationist position, including such fanciful ideas as that of the homunculus, is unscientific. Yet subsequent scientific discoveries have not resulted in a hands-down victory for epigenesis. In fact, modern theories concerning DNA (deoxyribonucleic acid), the genetic material that contains the "blueprint" for each individual human, incorporate aspects of both positions.

Adherents of epigenesis were correct in maintaining that the sperm or embryo is not simply a smaller form of a baby or fully grown human. On the other hand, sperm cells and embryos *do* contain the model for the fully grown organism, in the form of DNA.

☐ THE EIGHTEENTH-CENTURY REVOLUTION IN HUMAN HEALTH

The term "revolution" has a number of meanings. It can, for instance, refer to an uprising—sometimes violent—that results in a change of governments. More generally, revolution can refer to a sudden and dramatic change. Judged in any terms, the eighteenth century (the 1700s) was a period of revolution. Politically this culminated in the French Revolution, which began with the overthrow of the monarchy in 1789, as well as the American Revolution of 1775 to 1783. Ironically, the American Revolution against the British was inspired in large part by the writings of a British philosopher, John Locke (1632–1704). Locke's generation had been inspired by events in England, which began with the violent English Revolution of 1642 and ultimately led to the establishment of much greater popular rule (government by the people) in that country.

Even more powerful than the spirit of political revolution sweeping Europe was the intellectual revolution in the arts, sciences, and philosophy. The era has been characterized as the Enlightenment, a period that saw the rejection of old beliefs and a new emphasis on the powers of the human mind to solve problems. The sciences reflected the Enlightenment spirit in a number of areas, as new discoveries revealed the unfolding complexities of the natural world.

Botanists of the 1700s made great strides in understanding photosynthesis (the process by which green plants absorb carbon dioxide and release oxygen in the presence of sunlight) and plant reproduction, but while these advances greatly influenced the life sciences, their social impact was overshadowed by changes in the world of medicine. New surgical techniques, as well as methods of disease treatment and prevention, greatly enhanced the quality of human life, creating a revolution whose benefits extend to the present day.

Life Science

ESSAYS

Words to Know

inoculation: The prevention of a disease by the introduction to the body, in small quantities, of the virus or other microorganism that causes the disease.

mortality: Death.

occupational health: An area of medicine concerned with the health hazards related to specific professions.

public health: A set of policies and methods for protecting and improving the health of a community through efforts that include disease prevention, health education, and sanitation.

scurvy: An illness caused by a lack of vitamin C that results in swollen joints, bleeding gums, loose teeth, and an inability to recover from wounds.

smallpox: A viral infection accompanied by fevers and chills, and characterized by the formation of a rash over large parts of the body. As the effects of the illness continue, the rash turns to pus-filled bumps or papules that, when infected, can cause death. Even those who survive, however, bear the scars left by the eruption of the papules.

Disease prevention

For centuries people had suffered and died from the effects of various diseases, and medicine seemed helpless to prevent this. At best, doctors could help provide relief from the symptoms, but the idea of actually preventing disease seemed impossible. At worst, medical treatments based on misinformation had tragic consequences.

The Enlightenment, however, saw the rise of an entirely new approach. With the spread of inoculation methods, beginning in England, it became possible for the first time to stop a disease such as smallpox even before it struck. At the same time, other physicians—also chiefly in England—worked to counter several of the worst occupational diseases of the era, among them the condition known as scurvy, a bone disease common among sailors on long voyages, and the "chimney sweep's cancer." Research into "chimney sweep's cancer" led to increased understanding of the ways in which exposure to certain substances by

Life Science

ESSAYS

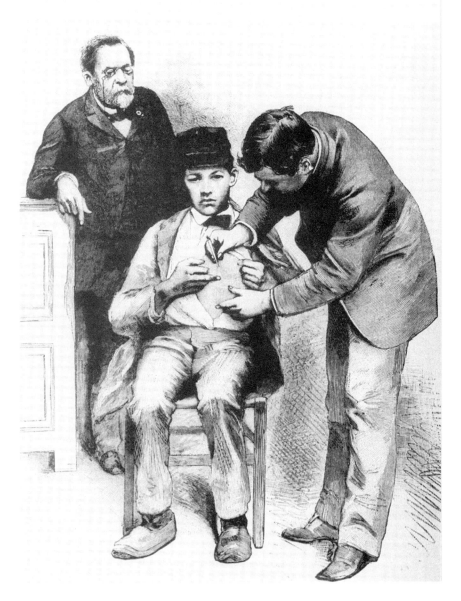

Louis Pasteur observes as a young boy receives an inoculation for rabies. With the spread of inoculation methods it became possible for the first time to stop a disease even before it struck. (National Institutes of Health/Corbis. Reproduced by permission of the Corbis Corporation.)

members of a particular profession can lead to increased mortality (death) rates.

Inoculation

Smallpox is characterized in its early stages by fever and by the formation of a rash. This rash eventually turns into pus-filled bumps, or papules, that may become infected, resulting in the death of the patient. Even smallpox survivors, however, bear scars on their skin that result from the eruption of the bumps.

Life Science

ESSAYS

About fifteen centuries earlier, doctors in India developed the principle of smallpox inoculation (see essay "'Alternative' Medicine Begins in the East" in this chapter), but European doctors were unaware of this when they began the first widespread efforts at inoculation during the 1700s. In any case, ordinary people as well as physicians had become aware of the fact that a person only became ill with smallpox once. This suggested that the introduction of a small quantity of smallpox to the body could actually prevent the patient from catching it later.

Scarred by a case of smallpox, Lady Mary Wortley Montagu (1689–1762) advocated the adoption in England of inoculation techniques then being practiced in Turkey. Her efforts were successful enough that inoculation spread to the American colonies, where Cotton Mather (1663–1728) became one of the leading advocates of inoculation. Ironically, Mather, a prominent Christian minister, met opposition from religious believers who maintained that the disease was a part of God's will.

People also feared inoculation because it seemed to make no sense: in order to be protected against the disease, the patient actually had to be partially infected with the virus that caused it. Indeed, some early patients did die, but thanks to the efforts of English physician Edward Jenner (1749–1823), inoculation became a common practice in England by the late 1700s. Instead of infecting patients with the more dangerous smallpox, he used cowpox, and he applied it in carefully measured doses, making inoculation much safer.

The principal of inoculation would in later years lead to the development of vaccines against a host of diseases, such as yellow fever and measles, that had previously claimed millions of lives and left millions more suffering painful aftereffects. As for smallpox, inoculations against it became so widespread that in 1978 the United Nations (UN) declared that it had been eliminated worldwide.

Scurvy
Sailors and ships' captains had long been aware of a disease that afflicted crews who had been at sea for long periods of time. Called scurvy, it could result in swollen joints, bleeding gums, loose teeth, and an inability to recover from wounds. Scientists today recognize scurvy as resulting from a deficiency of vitamin C, available in citrus fruits such as oranges. At the time, however, the concept of vitamins was unknown, and sailors at sea continued to live on a diet that consisted primarily of salted meats and hard biscuits—items that could easily be stored without spoilage in an era before refrigeration.

In the mid-1700s, Scottish physician James Lind (1716–1794) conducted a shipboard experiment in which he treated twelve sick sailors with six different remedies. Ensuring that all other factors in their diets remained the same, he gave two men the same treatment day after day and found that those given oranges and lemons recovered the most quickly. This was the first example of a clinical trial, or the testing of a treatment by careful and well-documented experimentation in which other variables or factors are unchanged.

It would be another half-century before the British navy adopted Lind's techniques. Another Scottish physician, Gilbert Blane (1749–1834), had long fought for the adoption of Lind's method, and finally in 1796 he convinced the navy to give each sailor a daily ration of lemons. At that time, the term "lime" was common for both lemons and limes, and as a result British sailors became known as "limeys." Eventually the treatment spread to the population as a whole, but outbreaks of scurvy still continued until after World War I (1914–18), when doctors isolated vitamin C as the controlling factor in scurvy prevention.

Life Science

ESSAYS

In the mid-1700s, Scottish physician James Lind discovered that eating oranges and lemons cured sailors of scurvy.
(© Orion Press/Corbis. Reproduced by permission of the Corbis Corporation.)

Chimney sweeps' cancer and occupational health

The eighteenth century marked the beginning of another revolution—the Industrial Revolution, a period of rapid development in technology marked by the introduction of powered machinery. The Industrial Revolution ultimately paved the way for the advances of modern society such as automobiles, computers, and other machines that have greatly improved human life, but it also had a number of negative consequences in its early stages. One of these was pollution, caused by the widespread use of coal to provide heat and power.

The chimneys of England continually became clogged with thick layers of soot, and someone had to clean them. Because the width of the chimneys was small, this job usually fell to young boys known as chimney sweeps. This fact alone illustrates the most cruel and inhumane aspects of the early Industrial Revolution: some of these boys, who were either abandoned children or members of desperately poor families, went to work at ages as young as four. By the age of seven, they were already too large to climb down the chimneys, which were so tight that the chimney sweeps

Life Science

ESSAYS

could wear no clothes while doing their job. As a result, soot became embedded in their skin, and since few people bathed, it stayed there.

In 1775 English surgeon Percivall Pott (1714–1788) described the high incidence of cancer in the scrotum (the sac containing the testicles) of former chimney sweeps. He related the disease to their occupation and concluded that their prolonged exposure to soot was the cause. His work sparked a series of reports by other scientists, and along with other famous figures such as author Charles Dickens (1812–1870; author of *David Copperfield* and *A Tale of Two Cities,* among others), he brought to light one of the most disgraceful facts of life in England. Yet it would be another century before child-labor laws forever put an end to the use of children as chimney sweeps.

Aside from the humanitarian aspect of Pott's work, which was motivated as much by compassion for the chimney sweeps as by a desire to discover the causes behind disease, his efforts increased awareness of *occupational health*. Occupational health remains an issue in modern society, and a number of laws in the United States and other countries help to protect workers in various professions from the hazards associated with their jobs.

Developments in public health
Aside from the occupational-health concerns raised by Pott and others, the eighteenth century saw a rising concern for *public health,* a set of policies and methods for protecting and improving the health of a community. Public health involves a number of components, including disease prevention, health education, and sanitation.

The most important European figure in the public health movement was Johann Peter Frank (1745–1821), who established the concept of "medical police" in the German state of Prussia. Frank's idea meant that government medical policy should be implemented through enforced regulations, much as governments have always upheld their laws through the use of police forces.

Among his concerns were sanitation, population control, and an early form of social welfare policy whereby new mothers were freed from the economic need to work outside the home. In France, physician Joseph-Ignace Guillotin (1738–1814), for whom the guillotine was named, also advocated a form of health police, and this led to the formation of a national health committee after the French Revolution (1789).

England and America
English public health efforts began with a campaign against alcoholism, leading to the passage of laws that controlled consumption. Advocates of

this policy pointed to a decline in death rates, in particular among infants, many of whom had been affected as fetuses by the alcohol consumption of their mothers.

Child mortality itself became a powerful public issue, and in 1741 the Foundling Hospital of London, England, was established to provide nursing and other care for orphaned and abandoned children. The work of Jenner, Lind, and others may also be seen as part of the trend toward increased awareness of public health in England.

The concept of public health took root early in the American colonies that became the United States. Concerns about diseases carried by ships' crews arriving in Boston Harbor, in Massachusetts, led to the formation of "the Selectmen," a group of physicians who inspected ships and crews for smallpox and other diseases. New York City meanwhile passed street-cleaning laws, and in the mid-1700s undertook inoculation efforts as well.

Life Science

ESSAYS

A satirical cartoon showing the Thames River and its offspring—cholera, scrofula, and diptheria. The Industrial Revolution paved the way for advanced modern society but it also brought negative consequences such as pollution. (Reproduced by permission of Hulton/Archive.)

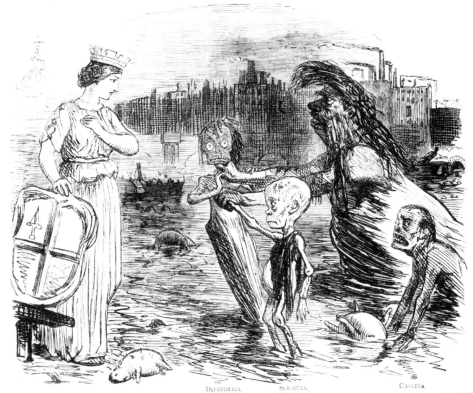

Life Science

ESSAYS

The growth of hospitals

Hospitals had existed in Europe and the Muslim world since the early Middle Ages (roughly 500 C.E. to 1500 C.E.), but the modern concept of the hospital did not appear until the 1700s. Only in that era did the hospital become a fundamental part of medical care and education, and within another century it would become the focal point of the medical profession.

This movement probably began at the University of Leiden in the Netherlands, where medical professor Hermann Boerhaave (1668–1738) expanded efforts to provide instruction for medical students by allowing them to conduct hands-on care of patients. Establishing a model still used by doctors, he took his students on rounds throughout the sick ward, allowing them to observe his interactions with patients and listen to his comments on various diseases and conditions exhibited by the sick.

At the time, such regular doctor–patient interactions were unusual, but Boerhaave's students soon spread his teaching methods to other European cities. Hospitals sprang up throughout Europe during the 1700s, and doctors increasingly focused their efforts around these large institutions. Hospital growth in the United States, by contrast, was much slower: in 1800, America had just two major hospitals (in Philadelphia and in New York) for a national population of five million.

Nevertheless, by the end of the eighteenth century models of the modern Western hospital were in place throughout Europe, and the idea caught on in the United States during the early years of the nineteenth century. Yet even in the 1800s, hospitals were perceived primarily as places of care for the poor, whereas wealthy individuals continued to receive care at home when possible. With the development of anesthesia and advanced surgical techniques during that century, however, most serious medical care began taking place within the walls of a hospital.

Surgery and dentistry

The 1700s were a time of tremendous surgical advances. Surgeons, particularly in England and France, developed new techniques to control bleeding, drain and close surgical wounds, and repair wounds resulting from explosives and gunshots. They also increased their abilities to repair malfunctions in the urinary system and intestinal tract, and to perform cesarean sections (the surgical removal of a baby from its mother's uterus). In addition, they learned how to remove cataracts from the eyes (a cataract is the clouding of the lens), to extract gallstones, and to remove organs.

The Spontaneous-Generation Debate

Despite the many scientific advances of the 1700s, scientists still found themselves confronted with age-old superstitions and fallacies. Among these was spontaneous generation, the belief that living organisms could originate from nonliving matter. It is easy to understand how spontaneous generation came to be accepted: if food is left in an empty room, it attracts rats, and from ancient times onward, people assumed that the food had *become* rats. This belief extended to other creatures as well: flies, for instance, were thought to spontaneously generate from manure.

Italian physician and poet Francesco Redi (1626–1697) was one of the first to question the spontaneous origin of living things. In 1668 he discovered that when adult flies are kept away from rotting meat, maggots did not develop. Thanks to Redi, by the 1700s the spontaneous-generation debate had been limited to questions regarding microscopic life-forms.

Interestingly, the most serious blow against spontaneous generation came as an indirect result of work intended to prove that it was true. In 1745 English microscopist John Tuberville Needham (1713–1781) showed that microscopic forms of life appeared even in flasks that had been sealed and heated. Later, Italian physiologist Lazzaro Spallanzani (1729–1799) attempted to replicate Needham's experiments. When he did so, he discovered that Needham had not heated the flasks long enough to kill all the microbes, some of which could endure higher temperatures. Despite Spallanzani's findings, belief in spontaneous generation continued into the 1800s, when experiments by French chemist and microbiologist Louis Pasteur (1822–1895) and Irish physicist John Tyndall (1820–1893) effectively disproved it.

Particularly important was the work of surgeons in the area of amputation, a common practice at a time when medicine had little means to stop infection. In 1718 French surgeon Jean Louis Petit (1674–1750) developed an effective tourniquet to stop bleeding during an amputation, a breakthrough that some consider the most important advance in surgery prior to the development of anesthesia. Surgeons greatly improved their ability to

Life Science

BIOGRAPHIES

provide quick and clean amputations and to close wounds by using adhesive tapes. (The concept of using sutures or stitches did not develop until later.)

Dentistry

A long list of physicians contributed to advances in surgery, which touched on virtually every part of the human body. Likewise, a number of figures aided developments in an area of medicine that had largely been ignored up to that time—dentistry. Yet one man stands above the rest, French dentist Pierre Fauchard (1678–1761). Fauchard initiated a large number of techniques still in use today.

Due to the many nerves in the mouth and gums, problems with the teeth are some of the most painful conditions known, which explains why even today, most people dread visiting the dentist. In the absence of proper dental care prior to the modern era, people's teeth simply fell out—or they had to be pulled, an experience that was quite painful. Fauchard, who has been called the "father of scientific dentistry," was the first to identify the conditions that cause toothaches, including tooth decay.

In *The Surgeon Dentist* (1728), Fauchard described dental anatomy, tooth decay, dental surgery, gum disease, and other aspects of dentistry. He also recommended treatments such as the use of lead to fill cavities, as well as oil of cloves and cinnamon to prevent infection. Because of the many tooth problems that had afflicted people for centuries, dentures (false teeth) had long since been invented, but Fauchard improved on these by creating dentures with springs and real teeth.

Thanks to Fauchard, advances in dentistry spread throughout Europe and ultimately to the United States. The identification of nitrous oxide, or laughing gas, by British chemist Joseph Priestly (1733–1804) proved to be an effective anesthetic, offering relief to patients undergoing dental surgery. It would still be many years before the modern concept of daily brushing and flossing, as well as regular checkups, emerged. Only in the late twentieth century did dentures become a rarity and bright, healthy teeth the norm.

◻ BIOGRAPHIES

◻ ARISTOTLE (384–322 B.C.E.)

Greek philosopher and scientist

The importance of the Greek philosopher Aristotle to human civilization is incalculable. With his teacher Plato (c. 428–348 B.C.E.), he was one of the two most significant thinkers of ancient Greece. Driven by an interest in classifying various life forms, he developed the principles of biology, and through his student Theophrastus (c. 372–c. 287 B.C.E.) exerted an impact on botany as well. During the Middle Ages (roughly 500 C.E. to 1500 C.E.) Aristotle's writings remained among the most significant texts in the life sciences.

Life Science

BIOGRAPHIES

Tutor to Alexander the Great

Born in the town of Stagira in Macedonia to the north of Greece, Aristotle was the son of a physician named Nicomachus. At the age of seventeen Aristotle went to study at Plato's school in Athens, Greece, but in spite of their close personal relationship, Aristotle rebelled against Plato's teachings. Eventually his philosophical system would rival Platonism as a means of interpreting the world.

After studying for twenty years in Athens, Aristotle served a leader named Hermeias in Asia Minor (modern Turkey) and married Hermeias's daughter Pythias when he was forty years old. Shortly afterward, however, Hermeias was assassinated, and Pythias died giving birth to a daughter. On the heels of these tragedies, the grieving philosopher received a request from Macedon's King Philip II (382–336 B.C.E.) to tutor his son Alexander—better known as Alexander the Great (c. 356–c. 323 B.C.E.). Aristotle taught Alexander for four years, then at Philip's request helped rebuild his hometown of Stagira, destroyed in a war.

Scientific studies at the Lyceum

In 334 B.C.E., at age fifty, Aristotle returned to Athens and established his own school, the Lyceum. During the years that followed, he did much of his most important work both as a philosopher and a scientist. He enlisted his students in all manner of scientific pursuits, from studying animals' organs to analyzing the political systems of various Greek cities.

Aristotle.
(Reproduced courtesy of the Library of Congress.)

These activities had in common the idea of classification, or the formation of general ideas from specific examples. Not only did he give shape to biology as a discipline, but Aristotle's ideas on various subjects—for instance, his classification of a dolphin as a mammal and not a fish—placed him well ahead of his time. With his students, he dissected hun-

Life Science
BIOGRAPHIES

Aristotle studying nature. (© Corbis. Reproduced by permission of the Corbis Corporation.)

dreds of animals and presented a system for classifying all life forms that remained more or less unchanged until the time of **Carolus Linnaeus** (1707-1778; see biography in this chapter).

Later life

Aristotle remained in Athens until 323 B.C.E., when he received word that Alexander had died in Babylon. Fearing that Alexander's foes might now seek him out in an effort to kill all persons associated with the late emperor, he moved to a nearby island, where he died in 322 B.C.E. Late in life he had become involved with a beautiful woman named Herpyllis, but when he died he requested that he be buried beside his first love, Pythias.

As a philosopher, Aristotle established the foundations for scientific inquiry. Whereas Plato believed that people form ideas because those ideas themselves are "real" in some sense, Aristotle held that general ideas are formed simply as a result of specific information and experiences. When people experience enough things that seem similar, he maintained, they classify these together—an idea essential to science.

AVERROËS (IBN RUSHD; 1126–1198)

Spanish Arab physician and philosopher

Ibn Rushd, known in the West as Averroës (uh-VEER-uh-weez), is famous for his writings on **Aristotle** (384–322 B.C.E.; see biography in this chapter), which became widely used texts until the 1500s. In addition to this body of work, which earned him the nickname "the Commentator," he wrote numerous texts on medicine, as well as works on law, philosophy, astronomy, and religion.

Commissioned to write commentaries on Aristotle

Spain in Averroës's time was controlled by Arab Muslims, and his family, who lived in the town of Cordoba, were important lawyers. As a young man, he traveled to Marrakesh, Morocco, where he met Abu Ya'qub Yusuf, son of the sultan, or king. Abu Ya'qub persuaded him to prepare a set of commentaries (guides) on Aristotle's works, and the result was a vast set of books that corrected many distortions made in interpretations of Aristotle by previous writers.

In addition to his commentaries on Aristotle, Averroës wrote more than twenty works on medicine, the most important of which was *Kulliyat* (General medicine), which built on ideas from another Arab physician and philosopher, **Avicenna** (980–1037; see biography in this chapter). *Kulliyat* covers topics such as the human organs and hygiene, as well as the prevention, diagnosis, and treatment of diseases.

As an astronomer, Averroës built on Aristotle's idea that each planet is attached to an invisible sphere or orb, and that these spin in concentric circles, or circles within circles. His other works include *Tahafut at-tahafut* (The incoherence of the incoherence), in which he maintained that subjects of religion can be approached with the use of human reason.

Serving the sultans

Abu Ya'qub became sultan in 1169 and appointed Averroës to a position as a religious judge in Spain, but in 1182 he summoned Averroës back to Marrakesh to serve as court physician. Even after Abu Ya'qub's death in 1184, Averroës stayed on to serve the new sultan. However, when Muslim Spain and Morocco went to war against Christian forces from Europe in 1195, he fell into disfavor because his ideas in *Tahafut at-tahafut* seemed to go against established Muslim doctrine.

For a time, he was banished, or forced to leave. However, after the Muslims gained victory, the sultan called him back and removed all official orders against him. Averroës died on December 10, 1198, in Marrakesh. In the years that followed, his work appeared in Latin translations in Europe,

Averroës
(© Archivo Iconografico, S.A./Corbis. Reproduced by permission of the Corbis Corporation.)

Life Science

BIOGRAPHIES

where they became highly popular. In fact, Averroës's commentaries on Aristotle were so clear that they soon gained favor over Aristotle's original work, and they remained influential textbooks for many centuries.

◻ AVICENNA (IBN SINA; 980–1037)
Persian physician and philosopher

Known in the West as Avicenna, the Persian thinker Ibn Sina was among the most influential figures in European philosophy and science during the late Middle Ages (the Middle Ages lasted from approximately 500 C.E. to 1500 C.E.). His *Canon of Medicine* became the single most important medical text in all of Europe and remained in use from the late 1100s to the late 1600s. He also played a significant role in reconciling science with the Muslim faith, an effort that later influenced European thinkers in their attempts to accommodate science to the teachings of Christianity.

A child prodigy

Born in Afshana in what is now Afghanistan, Avicenna was raised in Bukhara, now part of Uzbekistan. He displayed an early talent as a student, and at the age of ten he had already read the entire Koran, Islam's holy book. He gained other useful knowledge from an Indian teacher who exposed him to Indian principles of mathematics, including the numeral zero, first used by Hindu mathematicians.

Encouraged by his family, who placed a high value on study, Avicenna began his formal education at age sixteen and had soon mastered the available texts on medical theory, natural science, and philosophy. He then supplemented his studies by beginning a medical practice, in which he discovered a great deal more through experiment and observation.

Appointed as court physician to the sultan, or king, of Bukhara, Avicenna gained access to the Bukhara's library and by the age of eighteen had read all its books. This exposed him to the work of **Aristotle** (384–322 B.C.E.; see biography in this chapter), whose writings initially upset him because he found himself unable to reconcile the Greek philosopher's teachings with those of the Koran. One day, however, his reading of another Islamic scholar helped him unlock the seeming contradiction, and he was so overjoyed that he gave alms (money) to the poor in gratitude.

The *Canon of Medicine*

When Bukhara's sultan died, Avicenna was forced to wander, and the next years were hard ones. Finally, however, he settled in the town of Isfahan (in Iran), where he spent the remainder of his career in service to the emir, or

Life Science

BIOGRAPHIES

Avicenna. (Reproduced by permission of Parke, Davis and Co.)

ruler. During this time Avicenna wrote some one hundred books on a variety of subjects, including an encyclopedia that ran to nearly two dozen volumes.

Most important to the scientific world was his *al-Qanun fi al Tibb,* known in Europe as the *Canon of Medicine.* The *Canon* consisted of five books on subjects such as medical theory, simple drugs, and pathology (the causes of disease). It showed the influence not only of Aristotle, but of the ancient physicians **Hippocrates** (c. 460–c. 377 B.C.E.; see biography in this chapter) and **Galen** (129–c. 199; see biography in this chapter), as well.

During the latter part of his career, Avicenna went on a number of military expeditions with the emir and thus had an opportunity to conduct important studies of botany and zoology. Like many Muslims of his time, he owned slaves, one of whom turned against him when Avicenna was in his fifties. Hoping to steal his money, the slave put the drug opium into Avicenna's food. With his knowledge of medicine, Avicenna was able to treat himself and recover. The overdose weakened him, however, and in 1037 he had a relapse and died.

☐ GALEN (129–c. 199)

Greco-Roman physician

Galen was the primary authority on medicine throughout the Middle Ages (c. 500–c. 1500). He is particularly known for his contributions to physiology, the

Life Science

BIOGRAPHIES

scientific study of the functions, activities, and processes of living things. These contributions arose mainly from information he gathered during the many animal dissections he performed, and from his insights regarding the functions of, and relationships between, various organs. Nonetheless, he was wrong about many things, and because science in Europe progressed little during the thousand years that followed his death, these errors were compounded.

Physician to gladiators and emperors

Galen was born in the Greek city of Pergamum, in what is now Turkey. His father, an architect, mathematician, and philosopher, oversaw his early education, but at age fourteen Galen began four years of formal schooling in Pergamum. Later he studied at Smyrna in Greece and Alexandria in Egypt. During this time he became dissatisfied with the way medicine was being taught, so he began writing his own medical book, *On the Movements of the Heart and Lung,* for the education of his fellow students.

In 157 Galen moved back to Pergamum, where his work as a doctor gained so much attention that he was appointed chief physician to a troop of gladiators. Gladiators were slaves who fought in stadiums while spectators cheered, and Galen gained valuable insights while treating their massive injuries. A war in 162, however, interrupted gladiatorial competitions, so Galen moved to Rome.

In the capital of the Roman Empire, Galen continued his studies, often conducting public demonstrations of anatomy, and once again he gained attention. By 169 he had received an appointment from Emperor Marcus Aurelius (121–180) to serve as the ruler's personal physician, and after Aurelius's death Galen went on to serve two other emperors.

The good and the bad of Galen

Throughout his career, Galen produced more than five hundred books and articles on medicine and other subjects. He noted a difference between

Galen dissecting an animal. (Reproduced by permission of the U.S. National Library of Medicine.)

veins, which carry blood to the heart, and arteries, which carry blood from it; discovered that muscles work in pairs; demonstrated that an injury to the spinal cord can cause paralysis, or the inability to move one's limbs; and recognized that urine flows from the kidney to the bladder.

For all his achievements, Galen made a number of mistakes. One of these was embracing the doctrine of the four humors, the belief that human health is governed by a balance of four bodily fluids. **Hippocrates** (c. 460–c. 377 B.C.E.; see biography in this chapter) had also supported humoral theory, but it was Galen's writings that would have the most effect on the medieval adoption of this fateful idea. With regard to circulation, or the flow of blood, he also put forth a number of incorrect concepts, which were only corrected by British physician William Harvey (1578–1657) centuries later.

Despite these shortcomings, Galen is rightly recognized, along with Hippocrates, as one of the great physicians of ancient times. Many of his inaccuracies were based on the fact that prevailing religious beliefs prevented him from conducting dissections of human bodies; therefore he had to generalize his findings from studies of animal anatomy. Nor could he have known that the Roman Empire would decline after Aurelius's death, and that Western progress in medicine would come to a virtual standstill for over a thousand years. During that time, his work—both good and bad—would have an enormous influence.

◻ WILLIAM HARVEY (1578–1657)
English physician

As the first scientist to accurately describe the workings of the human circulatory system, Harvey established the modern understanding of physiology. He based his research on extensive experiments and observations of animals and humans, rejecting ideas that were not confirmed by experiments, and in the process he discovered facts that contradicted the teachings of **Galen** (129–c. 199; see biography in this chapter). Because he lived at a time when Galen's writings still had the status of medical scripture, this led to a great deal of criticism, but Harvey's views have prevailed.

Personal physician to the king
Harvey was born in Folkestone, England, the eldest son of Thomas Harvey, a well-respected merchant. He attended the King's Grammar School at Canterbury, where he established an academic record that gained him admission to Cambridge University. After receiving his bachelor of arts degree in 1597, Harvey entered another prestigious institution, the Uni-

Life Science

BIOGRAPHIES

Life Science
BIOGRAPHIES

versity of Padua in Italy, where he earned his doctorate in 1602. He gained certification as a physician in 1609.

While training as a physician, Harvey worked as an assistant surgeon at Saint Bartholomew's Hospital in London, England, and later became a professor of anatomy at the College of Physicians and Surgeons. Harvey's notes from this period indicate that he had already begun to develop the concepts that would later lead to his monumental discoveries regarding the nature of the circulatory system. He would spend forty years at the university, during which time he also maintained a private medical practice and served as personal physician to King Charles I (1600–1649).

Writings spark controversy

In 1628 Harvey published one of the most important books in medical history, *Exercitatio anatomica de motu cordis sanguinis in animalius* (On the movement of the heart and blood in animals). In it, he showed that blood flowed from the right ventricle (chamber) of the heart, through the pulmonary artery to the lungs, and then from the lungs to the right ventricle via the pulmonary veins.

In so doing, Harvey overturned the claims of Galen, who maintained that blood moved between the right and the left ventricle via invisible pores. He also showed that the heart, and not the liver (as Galen had maintained), was the central organ of the circulatory system. These findings shocked the medical community, and as a result, Harvey was subjected to enormous criticism by his colleagues. Nonetheless, he patiently answered their charges, often writing personal letters and articles to explain how he had arrived at his conclusions.

William Harvey. (Reproduced courtesy of the Library of Congress.)

The response to his work led to a decline in Harvey's medical practice, and after Charles was dethroned in the English Civil War of 1642 to 1648, Harvey retired to the country. He did, however, publish another important book, *De generatione animale* (1651), in which he coined the term *epigenesis* to describe the process whereby an embryo develops in the womb. Once again, he went against prevailing doctrines of the day, which held that embryos were preformed, miniature individuals within the egg. Thus to the end, Harvey remained a rebel against misguided traditions.

◻ HIPPOCRATES (C. 460–C. 377 B.C.E.)
Greek physician and philosopher

Life Science
BIOGRAPHIES

Known as the "father of medicine," Hippocrates was the first doctor to identify the causes of disease in nature, rather than as a result of punishment from the gods. In the more than sixty books attributed to him, he established careful methods for medical treatment, stressing diet, rest, and a clean environment. Some of his books may have been written by his large group of followers; this is almost certainly true of the Hippocratic oath, a code of conduct for physicians attributed to Hippocrates.

A teacher and writer

Born on the island of Cos off the coast of what is now Turkey, Hippocrates came from a family that claimed direct descent from the Greek god of medicine, Asclepias. After studying medicine from teachers on Cos, he traveled around Greece and other lands, lecturing on his principles, and died in Larissa on the eastern coast of Greece.

Hippocrates maintained that doctors should aid the body in healing itself, ideas expressed in the many works written either by him or by his followers. These writings addressed virtually every aspect of medicine known at the time, from anatomy to medical philosophy, and from pathology (the causes of disease) to pediatrics (medical treatment for children).

The four humors and the Hippocratic oath

Unfortunately, Hippocrates subscribed to the theory of the four humors, an idea developed by Greek philosopher Empedocles (c. 490–430 B.C.E.), which maintained that the body contains four humors, or bodily fluids: blood, phlegm, yellow bile, and black bile. Adopting humoral theory, Hippocrates taught that an imbalance of humors caused disease and ill health.

Hippocrates. Reproduced by permission of the British Museum.

Humoral theory complemented the Greek idea that there are four elements, or basic substances, in nature: earth, air, fire, and water. This, of course, is incorrect: today chemists recognize about ninety naturally occurring elements, such as carbon and hydrogen. Likewise humoral theory has no basis in fact, and indeed, many of the treatments based on it (such as bleeding patients to remove "excess" blood) were often harmful.

Life Science

BIOGRAPHIES

Of greater value was the Hippocratic oath, which established a code of conduct for physicians. Though attributed to Hippocrates, the oath was probably written by his followers; in any case, it established much-needed guidelines for the medical profession. Physicians today still swear by the Hippocratic oath, and copies of it hang on the walls of medical offices all over the world.

ANTONI VAN LEEUWENHOEK (1632–1723)
Dutch naturalist

As the first scientist to observe bacteria and protozoa (single-celled organisms) under a microscope, Antoni van Leeuwenhoek (LAY-vin-hook) is regarded as the "father of microbiology." Though he lacked formal training as a scientist, Leeuwenhoek's skill, diligence, and endless curiosity helped him to overcome his lack of education. Open-minded and resistant to the outdated traditions established by past scientists, he routinely shared the fruits of his research, and in the process he opened up the previously invisible world of microscopic life to the European scientific community.

An amateur scientist
Leeuwenhoek was born in Delft, Holland, on October 24, 1632. He came from a family of tradespeople, or skilled workers, rather than educated professionals, and for this reason he received no educational training beyond the equivalent of grade school. After working for a time as a shop apprentice, he returned in 1654 to Delft and spent the remainder of his life there.

During the years that followed, Leeuwenhoek worked variously as a fabric merchant, a surveyor, and a city official. The city official position he took in 1660 gave him a regular income and freed him to indulge his hobby of grinding glass lenses and using them to study tiny objects. In this he seems to have been inspired by English physicist Robert Hooke's (1635–1703) influential work *Micrographia* (1665), in which Hooke discussed his own observations with the microscope.

Though Leeuwenhoek's microscopes were simple in design, he possessed tremendous skill in grinding lenses. He did not invent the microscope, which had been in use for some years, but he did develop the best lenses available at that time, and thus he is credited with greatly improving the instrument. As for the studies he conducted using it, he approached these with the attitude of a talented amateur, and thus his reports on his research lacked the organizational qualities they might have had if he had been a trained scientist.

Amazing discoveries

Nonetheless, Leeuwenhoek had an enormous capacity for careful observation, and by 1674 he had seen something no human eye had ever glimpsed: tiny, single-celled organisms that he dubbed *animalcules*. He also observed, and became the first to describe, spermatozoa (sperm), and later demonstrated that spermatozoa were necessary for fertilization of an egg. In addition, Leeuwenhoek's studies led to detailed descriptions of microscopic structures in muscle tissue, insects, plants, and yeast.

Of particular importance were Leeuwenhoek's findings with regard to spontaneous generation—the belief, still common at the time, that living creatures such as rats could be generated from nonliving items such as food. Using his microscope, he showed that both grain weevils and fleas were produced, not from grain and other substances that they infested, but through the same reproductive processes that produce other insects.

In 1680 Leeuwenhoek was elected to membership in the Royal Society of England, an extremely prestigious community of scientists. The Royal Society of England made many of his findings public, and the untrained son of a Dutch tradesman became a scientific celebrity in Europe. He gladly demonstrated his microscope for dignitaries the world over and continued to do so until his death in 1723.

CAROLUS LINNAEUS (1707–1778)
Swedish physician and botanist

Carolus Linnaeus established a system for classifying animal and plant life, still in use today, that greatly advanced biological studies. Under this system, animals and plants have unique and universally accepted two-part names (binomial nomenclature) that identify them by genus and species. The system also includes higher levels of organization, including phylum, subphylum, class, order, and family.

Not an outstanding student

The name Carolus Linnaeus is actually a Latinized version of the name with which the great scientist was born on May 23, 1707, in Råshult, Swe-

Antoni van Leeuwenhoek. (Reproduced by permission of Archive Photos, Inc.)

Life Science
BIOGRAPHIES

den: Carl von Linné. His father, Nils Linné, was a pastor and hoped his son would follow in his footsteps, but the boy showed more interest in his father's hobby of gardening. From an early age, Linnaeus decided that he would become a botanist.

Despite his later achievements, Linnaeus was not an outstanding student. Nonetheless, he entered medical school at the University of Lund in 1727 but transferred a year later to the University of Uppsala, where professors encouraged him to pursue his interests in botany.

Linnaeus accepted an appointment as lecturer at Uppsala in 1730, and two years later he received an appointment to survey Arctic plants and animals in Lapland, a region in northern Scandinavia. During his five months in Lapland, he wrote detailed accounts of his observations, which he published as *Flora Lapponica* in 1737.

Develops his classification system

On the heels of his work in Lapland, Linnaeus received invitations to present lectures and conduct other surveys; meanwhile, he began work on his classification system for plants. He tried at first to give each plant a single name but soon realized that two names were required. He did this for the same reason that people have two names: there may be plenty of men named John, but the number of John Smiths is much smaller. In the case of Linnaeus's binomial nomenclature, the name given to each species was unique to that species, while the genus or generic name identified groups of species with related characteristics.

Carolus Linnaeus.
(Reproduced by permission of the U.S. National Library of Medicine.)

Linnaeus published the first edition of *Systema naturae,* which introduced his classification system, in 1735. Also in that year, he received a medical degree from the University of Harderwijk in Holland, but within four years he had moved to Stockholm, the capital of Sweden. There he married Sara Moraea, whom he had met while on a survey in 1734. Appointed professor at Uppsala in 1741, Linnaeus combined his duties at the university with a role as chief royal physician, a position to which he was appointed in 1747. In 1758 the king of Sweden knighted him.

During this time, he continued to refine his classification system, and in 1758 he published his most influential work, the tenth (and much expanded) edition of *Systema naturae*. He returned to his original name of

Carl von Linnè and continued his work until his retirement in 1776. In 1774 he suffered a stroke and a second one in 1778 led to his death on January 10. He is buried in Uppsala and was honored in 1788 by the founding of the Linnaean Society in London, England. By the early twentieth century, the international community of zoologists and botanists had officially adopted Linnaeus's classification system.

◻ PARACELSUS (1493–1541)

Swiss physician, pharmacologist, and alchemist

Not only was Paracelsus the most important figure of Renaissance medicine, he was also one of its most colorful. On the one hand, he challenged the ineffective traditions then prevailing in European science and helped push medicine toward the modern era. On the other hand, he remained fascinated with alchemy, which involved unscientific concepts such as the belief that ordinary metals can be turned into gold. A maverick by the standards of any age, Paracelsus greatly improved Renaissance medicine by advocating chemical remedies rather than the herbal treatments that had prevailed throughout the Middle Ages (roughly 500 C.E. to 1500 C.E.).

Paracelsus. (Reproduced by permission of the U.S. National Library of Medicine.)

Eager for adventure

The name "Paracelsus" was a pseudonym, which he created by combining the Greek prefix *para-* (beside or beyond) with the name of the great Roman medical writer Celsus (flourished in the first century C.E.). His real name was Philippus Aureolus Theophrastus Bombast von Hohenheim, and that alone says a great deal about the outlandish figure who would one day shake the foundations of the European medical establishment.

Born in Einsiedeln, Switzerland, Paracelsus was the only son of Wilhelm von Hohenheim, a poor country physician. After his mother died, he moved with his father in 1502 to Villach, Austria, where his father taught chemistry. Young Theophrastus, as he was called, served as an apprentice in medicine, working with his father, before beginning a life of wandering in 1507.

Eager for both knowledge and adventure, he traveled widely and briefly attended several universities in Germany. He may have received his

Life Science

BIOGRAPHIES

bachelor's degree from the University of Vienna in 1510, and by 1513 he had enrolled at the University of Ferrara in Italy, where he apparently earned his medical degree in 1515. In any case, Paracelsus was never comfortable in a university environment and maintained that the true student should seek knowledge from sorcerers, nomads, thieves, and peasants. In time his journeys took him to England, and even to Africa and Asia.

Challenging the old masters

Despite his disdain for university life, by 1527 Paracelsus was lecturing at the University of Basel in Switzerland. He inaugurated his tenure in true Paracelsian fashion, publicly burning the works of **Galen** (129–c. 199; see biography in this chapter) and **Avicenna** (980–1037; see biography in this chapter). These two men, whose writings had defined medieval medicine, were so esteemed that Paracelsus could hardly have caused more commotion if he had burned a Bible.

Galen had taught that the four humors, or bodily fluids, govern human health, but Paracelsus rightly dismissed this as nonsense. Instead, he attributed the onset of disease to environmental factors, including contagion (the spread of a disease from one person to another), geographic location, and the effects of certain chemicals.

Chemicals, in fact, were a major theme of his work: from his experiments with alchemy, he devised the idea, later known as iatrochemistry, of using chemical treatments for illness. Thus he served as a critical link between unscientific alchemy and modern chemistry, chemotherapy (medical treatment with chemicals), and pharmacology (the science of drugs).

A trail of manuscripts

Everywhere he went, Paracelsus left manuscripts behind, but few of his writings were published in his lifetime. One of the exceptions was *Grosse Wund Artzney* (Great surgery; 1536), translated into Latin as *Chirurgia magna*. (It was typical of the unorthodox Paracelsus that he wrote in his native German rather than in Latin, the accepted means of communication for scientists of the time.) In *Chirurgia magna,* he analyzed gunshot wounds—a frequent problem in the war-torn years of the Renaissance—and argued against the then-common practice of using hot oil as treatment.

In other writings Paracelsus discussed syphilis, which had broken out in Europe late in the fifteenth century; drug therapy; occupational health hazards (his writings on diseases of miners was one of the first books on occupational health); and mental illness. One of the first scientists to link mental health with physical conditions, he rejected the popular notion that insanity was caused by demons.

Throughout his career, Paracelsus remained a medical outlaw, a fact that had as much to do with his coarse personality as with his unconventional views. He carried a sword with him wherever he went, a habit he adopted in 1524 and continued until his death. He died in Salzburg, Austria, in 1541, apparently as the result of a fight in a bar.

◻ RHAZES (AR-RAZI; c. 865–c. 923)
Persian physician

An individual of broad interests, Rhazes wrote a vast encyclopedia of medicine that remained a highly significant textbook in the West for centuries. He also left behind a pioneering work on smallpox, as well as some two hundred other books and papers. As the first physician-in-chief for the prestigious Audidi hospital in Baghdad, today the capital of Iraq, he compiled a number of useful case histories that were used to instruct medical students for years to come.

Physician-in-chief of Baghdad hospital

Rhazes's actual name was ar-Razi; like most scientific figures of medieval Islam, he became known in the West by a Latinized version of his Arabic name. Born near present-day Tehran, Iran, he showed an early interest and ability in a wide range of fields that included music, alchemy, philosophy, mathematics, and physics. Medicine, however, attracted his greatest interest, and he resolved to become a physician.

Not only did Rhazes practice as a doctor in the years that followed, but he also devoted himself to continual study. He compiled accounts of medical treatments from a number of sources, tested them himself, added his own observations and improvements, and wrote works outlining his results. Because of his work he became widely known and respected and was chosen as physician to the court of Adhud Daulah, the ruler of Persia (Iran).

Later the caliph, the ruler of the entire Islamic world, commissioned him to establish and direct the new hospital in Baghdad, capital of the Muslim empires. Rhazes selected the site for the hospital, the Audidi, by hanging pieces of meat at various sites around the city. He chose the area where the meat spoiled most slowly on the theory that the air quality was best there.

Rhazes.
(Reproduced by permission of the Granger Collection, New York.)

Life Science

BIOGRAPHIES

Later he set up a special section in the hospital for patients with mental illnesses, making this perhaps the first real psychiatric ward, as opposed to the prison-like asylums that would spring up in Europe a few centuries later.

Extensive medical work and writing

Rhazes had modern ideas in other areas as well. He tested the effects of experimental therapies on animals before administering them to humans and used opium, a narcotic drug derived from the juice of the opium poppy, as an anesthetic during surgery. In addition, he was the first doctor to use alcohol for medicinal purposes. He also instituted a set of professional standards for the practice of medicine that emphasized the humane treatment of patients and dabbled in the fields of obstetrics, gynecology, pediatrics, and ophthalmology—the branches of medicine respectively concerned with childbirth, the female reproductive system, the care of children, and the eyes. He gave the first description of the surgical procedure for the removal of cataracts, a clouding of the lens. His book *al-Judari wa al-Hasabah* (A treatise on smallpox and measles), the first significant study of smallpox, went through forty editions and remained in use through the nineteenth century.

The most well-known among Rhazes' many works was *Kitab al-hawi* (Comprehensive book of medicine), a twenty-volume encyclopedia containing the details of Greek, Syrian, and Arab medical knowledge, combined with his own practical experience. Translated into Latin in 1279 and reprinted a number of times after the invention of the printing press, it was still being used in the 1500s. In addition to his extensive work as a doctor and a writer, Rhazes was also a philosopher and a chemist. In his work as a chemist, he discovered sulfuric acid.

ANDREAS VESALIUS (1514–1564)

Belgian anatomist, physician, and surgeon

With his *De humani corporis fabrica* (On the structure of the human body), Andreas Vesalius revolutionized physicians' views of anatomy. Though he began his career as an admirer of **Galen** (129–c. 199; see biography in this chapter), whose ideas still dominated European medicine in the 1500s, his firsthand studies of anatomy led him to become one of Galen's strongest critics. By daring to challenge the ancient master, he helped open the way for a greater understanding of the body and its structures.

A family of physicians

He was born Andries van Wesel (later he Latinized his name to Andreas Vesalius) in Brussels, today the capital of Belgium. Vesalius came from a long line of physicians who had served the Holy Roman emperors, rulers over much of what is now Germany and other parts of central Europe. In honor of the family name, Holy Roman Emperor Charles V (1519–1556) granted Vesalius's father a coat of arms that depicted three weasels.

In 1529 Vesalius entered the University of Louvain in Belgium. He studied there until 1533, then at the University of Paris, France, until 1536. In 1537 he received his medical degree from the University of Padua in Italy, where he went on to teach surgery for the next seven years. Also in 1537 he published a commentary on the works of **Rhazes** (c. 865–c. 923; see biography in this chapter).

Doubts about Galen

While studying in Paris, Vesalius first began performing dissections. In those days it was not common for a physician to personally cut open a human body; more often, doctors relied on assistants to do the work, but Vesalius insisted on taking a hands-on approach. In fact, he became so fascinated with anatomy that he once stole the corpse of a criminal who had been executed.

While teaching at Padua, he undertook another then-unusual practice, hiring professional artists to prepare accurate anatomical charts for his students. In 1538 he published his first important work, *Tabulae anatomicae sex* (Six anatomical charts), which contained illustrations by the Flemish artist Jan Stephan van Calcar (c. 1499–c. 1546), a student of the great Italian painter Titian (c. 1488–1576).

The more he learned by direct observation, the more Vesalius came to doubt the principles of Galen. In 1540 at the University of Bologna in Italy, he publicly announced that Galen's ideas could not be applied to human anatomy because they were based not on studies of the human body—dissection of humans was forbidden in Galen's time—but on the anatomy of apes, pigs, and dogs. He went on to demonstrate his results in *De humani corporis fabrica*, complete with extensive illustrations by Calcar. These illustrations were so good that for many years people believed that Titian had actually painted them.

Later years

Vesalius personally gave a special copy of his monumental work to Charles V, who was so impressed that he appointed him as his household

Andreas Vesalius. (© Bettmann/Corbis. Reproduced by permission of the Corbis Corporation.)

Life Science

BIOGRAPHIES

Life Science

BRIEF BIOGRAPHIES

physician. In the years that followed, Vesalius turned from his work as an anatomist to concentrate on service to the king, but in 1546 he published an important work on syphilis, a sexually transmitted disease then ravaging Europe. In 1555 he published a new edition of *Fabrica,* but by then his student Gabriele Falloppio (1523–1562) had emerged as the preeminent anatomist in Europe.

In 1544 Vesalius married Anne van Hamme and spent much of his time thereafter traveling with the emperor throughout Europe, often serving as a military surgeon. When Charles V abdicated (abandoned the throne) in 1556, he gave Vesalius a noble title and a pension, or annual funding. Later Vesalius moved to Madrid, Spain, to serve as court physician to the Spanish king, and died in Greece while returning from a pilgrimage to Jerusalem.

BRIEF BIOGRAPHIES

ALBERTUS MAGNUS (c. 1200–1280)

Albertus Magnus was a German naturalist and philosopher whose detailed accounts of plants and animals in *De vegetabilibus et plantis* and *De animalibus* contributed to medieval understanding of botany and zoology. He also influenced the sciences in general by promoting new investigation rather than relying on the ideas of past scholars, as was the practice at the time. His work thus constitutes a turning point in the history of science.

ASCLEPIADES OF BITHYNIA (FLOURISHED IN THE SECOND AND FIRST CENTURIES B.C.E.)

Asclepiades was a Greek physician and philosopher noted for his rejection of the doctrine of the four humors. He also promoted the idea of the philosopher Democritus (c. 460–c. 370 B.C.E.) that matter is composed of atoms.

AVENZOAR (IBN ZUHR; c. 1091–1162)

Avenzoar, or Ibn Zuhr, was an Arab physician credited as the first parasitologist, a biologist who studies parasites. He also was the first to describe the skin condition known as scabies and provided the first clinical description of a tracheotomy, an operation in which a blockage is removed from the windpipe.

HERMANN BOERHAAVE (1668–1738)

Boerhaave was a Dutch physician and a professor of medicine who introduced the practice of teaching students by taking them on rounds through a hospital. He was also an early advocate of what might be called "bedside manner"—the idea that a physician should develop a friendly, trusting relationship with the patient. Although Boerhaave is not associated with any major discoveries in medicine or chemistry, he became so renowned as a medical teacher that a letter simply addressed "To the Greatest Physician in the World" was delivered to him.

◣ LOUISE BOURGEOIS (1563–1636)

Bourgeois was a French midwife who, in 1609, became the first woman to write a textbook on midwifery. A well-educated woman who was trained and licensed at the famous hospital Hôtel Dieu in Paris, France, she advanced the respectability and social standing of midwives. Later she became official midwife to the French court, as well as to Italy's powerful Medici family.

◣ ELIZABETH CELLIER (FLOURISHED c. 1680)

Cellier was an English midwife who established a corporation of midwives in London, England, in 1687, as well as a foundling hospital, or a hospital for orphans and abandoned children. Married to Frenchman Peter Cellier, she also unsuccessfully advocated a college for midwives.

◣ CELSUS (FLOURISHED IN THE FIRST CENTURY C.E.)

Celsus was a Roman medical writer known as the first important medical historian. In addition, he established the use of Latin as the language of science, establishing a tradition that would remain in effect for more than seventeen hundred years. His *De medicina* (On medicine) is universally regarded as an invaluable medical classic.

◣ PETER CHAMBERLEN THE ELDER (1560–1631)

Chamberlen was an English man-midwife and physician credited with inventing forceps, a tong-like instrument for extracting a baby during difficult births. For years, the Chamberlen family kept the specifics of the invention a secret, but eventually the use of forceps spread to the medical community at large.

◣ CHEN CHUAN (FLOURISHED ABOUT 700)

Chen Chuan was a Chinese physician who provided one of the first clinical descriptions of a key side effect of diabetes mellitus. Diabetes mellitus is a disease of the pancreas that causes an inability to process sugar, resulting in lethargy (chronic tiredness) and other conditions. Chen Chuan is noted for describing the sweetness of urine in patients suffering from that condition.

Life Science

BRIEF BIOGRAPHIES

▲ CONSTANTINE THE AFRICAN (c. 1020–1087)

Constantine was a medical scholar from north Africa (modern-day Tunisia) who began the practice of translating Arabic medical texts into Latin. He studied at the famous medical school in Salerno, Italy, before entering the monastery of Monte Cassino, where he translated some thirty-seven books, including works by **Hippocrates** and **Galen** (see biographies in this chapter).

▲ RENÉ DESCARTES (1596–1650)

Descartes was a French philosopher, physiologist, and mathematician. Credited as the "father of modern philosophy," he promoted the idea, known as iatromechanism or iatrophysics, that the human body can be understood as though it were a type of machine. He also maintained that the soul is located in the pineal gland of the brain. (See full biography in Math chapter.)

▲ PEDANIUS DIOSCORIDES (c. 40–c. 90)

Dioscorides was a Greek physician and pharmacologist of the Roman Empire who wrote one of the first important herbals, or guides to herbal medicine. As such he is regarded as the founder of Western pharmacology, or treatment using drugs. In *De materia medica* (The materials of medicine), he described some six hundred plant species.

▲ EMPEDOCLES (c. 490–430 B.C.E.)

Empedocles was a Greek philosopher who originated the doctrine of the four humors. He taught that there are four basic elements in nature (fire, air, water, and earth), and that aspects of these are reflected in the four humors, or bodily fluids, that make up the body (blood, phlegm, yellow bile or choler, and black bile). His ideas, as unscientific as they were, would have an enormous influence on **Hippocrates** and **Galen**, (see biographies in this chapter) and hence on all of European science for centuries.

▲ ERASISTRATUS (FLOURISHED c. 250 B.C.E.)

Erasistratus (air-uh-SIS-truh-tus) was a Greek physician and anatomist often credited as the "father of physiology." (Later, that title became more closely associated with **Galen** [see biography in this chapter].) Erasistratus not only studied the structure of the body but also tried to explain the functions of its parts, including the nerves and the heart valves. However, he was on less-scientific footing when he promoted the philosophy of pneumatism, or the idea that life is dependant on certain vapors or "spirits" in the body.

James Derham

James Derham (born 1762) was the first prominent African American physician and became recognized as a specialist on throat disorders. Born a slave, he served a succession of three masters, all of whom happened to be doctors who educated him in the medical profession. Under the encouragement of his third owner, Dr. Robert Love, he saved enough money by working as a medical assistant and apothecary (pharmacist) to buy his freedom in 1783. Soon afterward, he opened a medical practice in New Orleans, Louisiana.

While visiting Philadelphia, Pennsylvania, Derham met Benjamin Rush (1745–1813), the most prominent American physician of his day. Like others before him, Rush took an interest in Derham and persuaded him to move his practice to Philadelphia. Derham gained great respect in that city's medical community and became widely known as an authority on the relationship between disease and climate.

BARTOLOMMEO EUSTACHIO (1520–1574)

Eustachio was an Italian anatomist best known for identifying the part of the ear today known as the Eustachian tube. In addition, he wrote highly detailed treatises on the kidney, veins, and teeth, and with the help of an artist, he produced a set of magnificent engravings that greatly enhanced his work.

GABRIELE FALLOPPIO (1523–1562)

Falloppio, an Italian anatomist, is regarded—along with his teacher, **Andreas Vesalius** (see biography in this chapter)—as one of the founders of modern anatomy. He is best remembered for his identification of tubes that carry a fertilized egg from a woman's ovary to the uterus, structures that are today known as fallopian tubes.

PIERRE FAUCHARD (1678–1761)

Fauchard was a French dentist credited as the "father of scientific dentistry." In *The Surgeon Dentist* (1728) he described a number of aspects of dentistry, including tooth decay. He was the first to use gold inlays to fill a tooth after performing a root canal, and his inventions included improved dentures, as well as the dental drill.

◭ GIROLAMO FRACASTORO (c. 1478–1553)

Fracastoro was an Italian physician important for his work with regard to diseases and their causes. He gave a name to the disease of syphilis, which broke out in Europe during his lifetime. In addition, he promoted an early model of contagion, or the idea that diseases can be passed from person to person.

◭ JOHANN PETER FRANK (1745–1821)

Frank was a German physician credited as a pioneer of modern public health. His *System of a Complete Medical Policy* (1779–1827), considered the first systematic treatment of public health and hygiene, provided detailed models for hygiene laws. Among the measures Frank advocated was the establishment of "medical police" to protect people's health, an idea that became law in the German state of Prussia.

◭ LUIGI GALVANI (1737–1798)

Galvani was an Italian physician and pioneer of electrophysiology, the area of medical science concerned with the electrical component of the nerve system. After an accidental discovery of the relationship between electricity and muscle movement while dissecting a frog, Galvani proposed a theory of "animal electricity." He was incorrect in supposing that the electricity in living creatures is different from ordinary electricity; nonetheless, his research opened new lines of inquiry concerning the structure and the function of the nervous system in both animals and humans.

◭ GERARD OF CREMONA (1114–1187)

Gerard was an Italian scholar and translator notable for his role in bringing Arabic medical texts to the West. His most important work as a translator involved **Avicenna**'s (see biography in this chapter) *Canon of Medicine*, destined to become the most important work in the medical sciences of Europe for the next five hundred years. Gerard translated some seventy-five other Arabic texts, many of them on the sciences, into Latin as well.

◭ CONRAD GESNER (1516–1565)

Gesner was a Swiss physician, zoologist, and botanist who became one of the founders of modern zoology. His most important work was *Historiae animalium* (History of animals), published in five volumes from 1551 to 1587. The book, regarded as a major turning point in the history of zoolo-

Frederick II

Frederick II (1194–1250) ruled the Holy Roman Empire, which controlled much of what is now Germany, Italy, and other parts of Europe during the Middle Ages. A well-educated man, he took a profound interest in the sciences, but much of his time was also consumed with affairs of state, including his participation in the Sixth Crusade (1228–29). A number of laws established by Frederick advanced the legitimate practice of medicine and diminished the status of the many quacks who plagued European medicine at the time.

In 1221 Frederick ordered that no doctor could legally practice in the empire unless he had passed an examination by the authorities at the distinguished medical school in Salerno, Italy. Twenty years later, he became the first major European ruler to permit dissection of cadavers, formerly prohibited by religious law. Nor was he merely a spectator in the world of science. Frederick wrote a scholarly work, *On the Art of Hunting with Hawks*, that advanced medieval understanding of ornithology, the branch of zoology concerned with birds.

gy, provided a comprehensive catalog of all animals known to that time, along with over a thousand beautiful woodcut illustrations.

▲ REINIER DE GRAAF (1641–1673)

De Graaf was a Dutch anatomist who first identified the ovary. In his work on the female reproductive system, he established an important model of inquiry by not simply analyzing structures, but by attempting to develop an understanding of their functions. De Graaf is also noted for his role as illustrator of his own writing, for which he provided pictures that were far more detailed and precise than those that accompanied most medical texts of the day. Despite a very short career, he also contributed to understanding of the pancreas and its functions.

▲ GUY DE CHAULIAC (c. 1300–1368)

Guy was a French physician and surgeon who is often regarded as the leading medical mind of his day. Guy incorporated the strong emphasis on

Life Science

BRIEF BIOGRAPHIES

anatomy advocated by his teacher, French surgeon Henri de Mondeville (1260–1320), in his monumental work *Chirurgica magna* (Grand surgery). This work remained in standard use until the 1600s. Guy is also remembered for choosing to remain and treat patients when the Black Death (1347–51) reached the French town of Avignon, rather than fleeing to the safety of the countryside.

▲ STEPHEN HALES (1677–1761)

Hales was an English clergyman and physiologist who first suggested that electrical impulses in the body cause the muscles to contract. Though not formally trained in science, he made a number of contributions in several scientific fields. Hales is regarded as the "father of plant physiology," and conducted the most important studies in animal circulatory systems since those of **William Harvey** (see biography in this chapter) more than a century earlier. His experiments with gases would greatly influence the work of eighteenth-century chemists and physicists.

▲ HALY ABBAS (DIED 994)

Haly Abbas, or Ali ibn Abbas al-Majusi, was a Persian physician whose *al-Maliki* (*The Royal Book*) was one of the first important medical encyclopedias of the Muslim and European worlds. *The Royal Book* discusses topics ranging from anatomy to drugs and was the first Arabic work to give detailed instructions regarding surgery. In writing it, Haly Abbas attempted to correct errors of both **Galen** and **Rhazes** (see biographies in this chapter), whose *Comprehensive Book* was, in his opinion, too long and poorly organized.

▲ JOHN HUNTER (1728–1793)

Hunter, a Scottish surgeon, is sometimes credited as the "father of modern surgery." In contrast to the then-common practice of debridement, or opening a wound to promote drainage, Hunter advocated allowing a wound to heal "under a scab." One of his most extraordinary acts was to infect himself with syphilis in an effort to demonstrate that it and gonorrhea (another sexually transmitted disease) were simply different manifestations of a single condition. He suffered and was weakened by this experiment, but survived. Though uneducated in his youth, Hunter surpassed the achievements of his older brother William (1718–1783), a distinguished London surgeon, who later became extremely jealous of his brother's stature in the medical community.

▲ IBN AN-NAFIS (DIED 1288)

Ibn an-Nafis was an Egyptian physician who first described pulmonary circulation, or the flow of blood between the heart and the lungs. In this

Hildegard von Bingen

Hildegard von Bingen (1098–1179) was a German medical writer and the most important woman in the life sciences during the Middle Ages (roughly 500 C.E. to 1500 C.E.). Her text *The Book of Simple Medicine* was among the most influential herbals, or catalogues of plants and their medical properties, during the era. But that was not the full extent of the achievements made by Hildegard in an age when members of her gender were virtually excluded from any meaningful participation in public life.

When she was eight years old, Hildegard's family dedicated her to the church, and she entered a convent near the German town of Bingen. There she received an education in Latin and other subjects, and she later achieved the position of head teacher. Five years later, encouraged by Bernard of Clairvaux (1090–1153), one of the leading figures in the church, she began writing about mystical visions she had been experiencing since childhood.

In contrast to her mystical side, which also found expression in the plays and the music she composed, Hildegard had a strong practical interest in herbal medicine. At the convent, she cared for the sick and acquired the knowledge that she put into her writings. Here she discussed the causes, diagnosis, and treatment of diseases, as well as subjects that included psychology, physiology, and human sexuality.

Much admired by her contemporaries, Hildegard became the first woman authorized by the pope to write works on theology, the branch of philosophy concerned with the nature of the relationship between God and humans. She was also the only medieval woman permitted to preach publicly. At fifty, already an advanced age by medieval standards, she established a new convent in Bingen and continued to work, write, and travel tirelessly until her death thirty-one years later.

he was many centuries ahead of **William Harvey**, (see biography in this chapter) but his findings did not become widely known until the twentieth century.

Life Science

BRIEF BIOGRAPHIES

▲ JAN INGENHOUSZ (1730–1799)

Ingenhousz was a Dutch-British plant physiologist best known for his discovery of photosynthesis, the process by which green plants absorb carbon dioxide and release oxygen in the presence of sunlight. He showed that this process takes place almost entirely in the leaves, as the result of exposure to sunlight rather than to heat.

▲ EDWARD JENNER (1749–1823)

Jenner was an English physician who developed the first workable modern form of inoculation, the protection of a population against a disease by carefully controlled treatments involving use of the microorganisms that cause that disease. During the late 1700s, Jenner inoculated British patients against smallpox and developed a method of ensuring the administration of exact doses.

▲ LEONARDO DA VINCI (1452–1519)

Leonardo was an Italian artist and scientist whose imagination, curiosity, and talent represent all that was best about the Renaissance (the period of artistic and intellectual rebirth, lasting roughly from 1350 to 1600, that marked the end of the Middle Ages). Famous for paintings such as the *Mona Lisa* and *The Last Supper,* his interest in representing the human form accurately led him into extensive anatomical investigation. He may have dissected as many as thirty human bodies, but he did not complete an ambitious project to create an atlas of human physiology. Most likely he did not have time: in addition to his work as an artist and anatomist, Leonardo was a skilled engineer, inventor, and architect. (See full biography in Technology and Invention chapter.)

▲ LI SHIH CHEN (1518–1593)

Li Shih Chen was a Chinese pharmacologist and physician who wrote a comprehensive and significant encyclopedia of Chinese medicine. *Pen ts'ao kang mu* (Great herbal; 1578) contained descriptions of more than eighteen hundred drugs and ten thousand prescriptions. Since Chinese medical practices tended to be regional, Li's compendium—which consisted of fifty-two volumes and took him twenty-seven years to complete—was particularly valuable as a guide to medicine throughout China.

▲ MARCELLO MALPIGHI (1628–1694)

Malpighi was an Italian physician and biologist who pioneered methods of studying living organisms with the aid of the then-recently invented microscope. He thereby founded the science of microscopic anatomy, and because of his work with tissues and cell samples he is also called the "father of histology" (the microscopic study of tissues). As the first scientist to demonstrate that capillaries connect small arteries and veins, he provided the factual data for **William Harvey**'s (see biography in this chapter) groundbreaking theory regarding the circulation of blood.

Life Science

BRIEF BIOGRAPHIES

▲ MONDINO DEI LIUCCI (c. 1270–c. 1326)

An Italian anatomist, Mondino was the first physician known to conduct a public dissection of a human body. In his time, dissections—long outlawed for religious reasons—had only recently become legal. Mondino's *Anothomia* (Anatomy; 1316) was the first real textbook on anatomy, and in part because of this, it remained the most important book in the discipline for more than two centuries, until *De humani corporis fabrica* by **Andreas Vesalius** (see biography in this chapter).

▲ MARY WORTLEY MONTAGU (1689–1762)

Montagu was an English writer and activist whose efforts led to the introduction of smallpox inoculations to England. Born into wealth as a member of the nobility, as a child Montagu suffered smallpox, which scarred her face. While in Turkey with her husband Edward, an ambassador, she became aware of local medical practices and arranged for her son to be inoculated against smallpox. Returning home, Montagu initiated smallpox inoculations in England and earned praise for preventing thousands of deaths annually.

▲ AMBROISE PARÉ (1510–1590)

Paré was a French surgeon who invented or introduced a number of surgical instruments and techniques. He is regarded as the "father of modern military surgery," and as the greatest military surgeon prior to the time of his countryman Dominique-Jean Larrey (1766–1842). Paré became highly critical of the practice adopted from the Arabs, and in use for over five hundred years by that time, of cauterizing (searing) gunshot wounds; instead, he advocated using bandages with treatments such as rose oil, egg yolk, and turpentine. These, he found, caused the wound to heal much more quickly and cleanly. In his writings, Paré addressed areas that included obstetrics, presenting a technique for delivering a breech (feet-first) baby that would not be improved upon until the 1800s.

▲ PAUL OF AEGINA (c. 625–c. 690)

Paul was a Byzantine (Greek) physician and surgeon and a leading authority on gynecology and obstetrics from ancient times until the Renaissance.

Life Science

BRIEF BIOGRAPHIES

Paul knew so much about these subjects—the branches of medical science concerned, respectively, with the female reproductive system and with childbirth—that he gained the respect of midwives and became recognized as an early "man-midwife." His writings showed a broad knowledge of medicine and in fact contained almost everything known on the subject in the West at that time. Later, his work had an enormous influence on Arab medical practice.

▲ PHILIPPE PINEL (1745–1826)

Pinel was a French physician who advocated the humane treatment of the mentally ill. As chief physician at the Paris asylum for men, Bicêtre, he made his first significant reform by unchaining patients, many of whom had been restrained for as many as forty years. He did the same for the female inmates of the Salpêtrière asylum after becoming director there in 1794. Pinel argued that diseases must be explained in terms of the findings revealed in an autopsy, rather than by reference to outmoded ideas based on the still-lingering remnants of humoral theory. These ideas had an enormous influence on French scientist Marie-François-Xavier Bichat (1771–1802), who developed a workable theory of human tissue.

▲ PERCIVALL POTT (1714–1788)

Pott was an English physician and surgeon. He was a pioneer in the field of occupational health by drawing public attention to a form of cancer that affected England's chimney sweeps. This disease came to be known as Pott's puffy tumor and was one of several conditions named for Pott, who worked in a number of medical fields. Nonetheless, his role in identifying and drawing attention to Pott's puffy tumor was his most important accomplishment. By publicizing the fact that boys as young as four were crawling through chimney pipes—thus gaining exposure to elements that would cause cancer twenty or more years later—Pott helped bring about child-labor legislation that ended this brutal practice.

▲ JOHN RAY (1627–1705)

Ray was a British botanist and zoologist who classified flowering and nonflowering plants extensively, thus laying the groundwork for the system developed later by **Carolus Linnaeus** (see biography in this chapter). Often referred to as the founder of British natural history (the study of developments in nature), Ray—who was also a minister—advocated natural theology, or the doctrine that the study of nature revealed God's wisdom, goodness, and power. Ray and English naturalist Francis Willughby (1635–1672) undertook the huge task of documenting the complete nat-

Franz Anton Mesmer

Swiss physician Franz Anton Mesmer (1734–1815) is known for Mesmerism, his theory of animal magnetism that addressed the functions of the body and the soul. He became so influential in his time that the verb "mesmerize," meaning to hypnotize or mentally overwhelm someone, is still in use—yet Mesmerism itself has long since been discarded as an unscientific form of medicine.

Mesmer argued that a force generated by the interactions of particles in the body existed, which turned the body into a sort of highly energized battery. Disease, he maintained, was caused by a blockage in the flow of this energy. Mesmeric treatment of patients involved the use of magnets, as well as the laying on of hands by a practitioner, in order to impart some of his energy to the patient. (Laying on of hands refers to touching someone as a means of healing that person, usually through spiritual means.)

The historical origins of hypnotism (or "artificial somnambulism," as it was called then) came from Mesmerism and probably explains why people got such amazing results from mesmeric treatment: because they *believed* they would. To Mesmer, hypnotism proved that the energy forces in all living things were linked. Later, members of the Romantic literary and artistic movement latched on to this idea as well.

In fact, Mesmerism was much more successful in the world of art than in that of science. Due to scandals surrounding mesmeric treatment, the French government undertook an investigation of Mesmerism in 1784 and concluded that it was a dangerous form of quackery. By then Mesmer, after living in Switzerland and Austria, had relocated to France, where he enjoyed his greatest success—and developed a questionable reputation for his influence on female patients.

History has not looked kindly on Mesmerism, but in the 1700s it seemed like a highly advanced means of explaining the soul in terms of science rather than of religion. Even today, the effects of Mesmerism linger in psychologists' use of hypnotism to treat certain conditions and in the idea—common to chiropractic and other forms of alternative medicine—that fields of energy run through the body.

Life Science

BRIEF BIOGRAPHIES

ural history of all living things, with Ray responsible for the plant kingdom and Willughby responsible for the animal kingdom. Ray's insight that fossils were the remains of living organisms was a significant advance over most other theories of his time.

▲ RENÉ-ANTOINE FERCHAULT DE RÉAUMUR (1683–1757)

Réaumur was a French naturalist, physiologist, and physicist who pioneered in the field of entomology, or the study of insects. Among his achievements were the development of a new classification system based on the insect's behavior rather than merely its body type; the identification of the queen bee and her role; and studies in bee communication. He also developed a new temperature scale, intended to resolve shortcomings of the Fahrenheit scale. His Réaumur scale had the advantage of placing zero degrees as the freezing point of water, but it was later replaced by the Celsius scale, which set water's boiling point at one hundred degrees instead of eighty degrees, as in the Réaumur system.

▲ LAZZARO SPALLANZANI (1729–1799)

Spallanzani was an Italian physiologist whose work helped discredit spontaneous generation, the belief that living organisms could originate from nonliving matter. A creative and extremely curious researcher, he approached subjects in biology that ranged from sexual reproduction to blood pressure to the navigation devices of bats. He also ventured into the physical sciences, most notably by studying lava flows inside an active volcano.

▲ JAN SWAMMERDAM (1637–1680)

Swammerdam was a Dutch anatomist and one of the leading pioneers in the use of the microscope for biological study. Unlike most early microscopists, who focused on plant tissue, Swammerdam's primary interest was insects. Indeed, he and René-Antoine Ferchault de Réaumur (1683–1757) are regarded as the founders of entomology, or the study of insects, and his *Biblia naturae* (Bible of nature, 1737) provided the first systematic account of insect anatomy, classification, and development. In addition, Swammerdam conducted dissections and microscopic examinations of human cadavers, which led to important discoveries about organs, most notably the lungs.

▲ THEOPHRASTUS (c. 372–c. 287 B.C.E.)

A student of **Aristotle**, (see biography in this chapter) Theophrastus is regarded as the "father of botany." Originally named Tyrtamus, he received the nickname Theophrastus ("divine speech") from Aristotle himself, who later appointed him his successor as director the Lyceum, a school in

Athens, Greece. An extremely prolific (active) writer, Theophrastus—in addition to numerous other writings—produced more than two hundred works on botany alone. Most important among these were *De Causis Plantarum* and *Inquiry into Plants,* in which Theophrastus outlined the basic concepts of plant classification and development. Theophrastus's work remained the most advanced in the field of botany for the next two thousand years.

Life Science

RESEARCH AND ACTIVITY IDEAS

▲ THOMAS WILLIS (1621–1675)

Willis was an English physician who conducted foundational research on the structure of the human central nervous system. A leading member of the iatrochemists, scientists who attempted to explain bodily functions and disease from a chemical standpoint, Willis influenced a distinguished group of men who studied under him: philosopher John Locke (1632–1704), British physicist Robert Hooke (1635–1703), and architect-scientist Christopher Wren (1632–1723). In 1664 Willis published his research on the central nervous system and various diseases in *Cerebri anatomi,* the most accurate account of the nervous system to date. In it, he described and explained the function of the circle of arteries located at the base of the brain, a complex of blood vessels still called the Circle of Willis.

RESEARCH AND ACTIVITY IDEAS

(1) Most medical and biological advances that affect life today originated in the West; however, as noted in "'Alternative' Medicine Begins in the East" and "The Great Doctors of Medieval Islam," doctors and scientists in other areas made a number of important advances. Research and prepare an essay on medicine in a part of the world other than Europe or countries such as the United States that have been strongly influenced by Europe. What is the status of medicine in those areas today? Do doctors use only Western forms of medicine, or do they use a combination of traditional and Western techniques? Also, consider the question of why Western progress in biology and medicine has received far more attention than the efforts of doctors and scientists in other societies.

(2) Illustrated books have long been an important element of the life sciences, especially after the invention of the printing press made possible the wide distribution of medical and biological texts. Form a team and create a small, illustrated text regarding some subject in the life sciences: for instance, varieties of plants or the human anato-

Life Science

RESEARCH AND ACTIVITY IDEAS

my. The team should elect a leader, who will assign jobs including research, writing, illustration, and book layout. The leader will oversee the editing of the "book," which can be photocopied for distribution to the class. At the end of the project, the class should review and evaluate each team's texts.

(3) Though women have a number of special medical needs, for centuries the medical establishment ignored them and largely prevented them from taking any medical profession other than that of midwife. Your job is to take a position regarding women's status in premodern medicine, and make a case for it in an essay and/or a class discussion. Explain either why women should not have been excluded, or (and this is more challenging to a modern person) pretend that you are a male physician of the era and explain reasons why they should have been.

(4) Imagine that you lived prior to about 1700, in a world that had no concept of preventive medicine, germs, the causes of diseases, anesthesia, or even bathing. Consider the effects that this would have on your body, as well as the things that people do today (for instance, brushing their teeth) to counteract those effects. As part of this activity, you might research and discuss aspects of premodern life sciences: medical treatments such as bloodletting or purging, or diseases such as the Black Death. Consider how you might respond to a disease such as the Black Death if you had no idea of the biological causes—for instance, would you pray? Would you blame others? Would you shun neighbors who had it?

(5) Carolus Linnaeus's system of scientific classification greatly advanced studies in the life sciences, and it is important to consider the standards he may have applied in grouping certain species together. Take a nature walk, visit a zoo, or review varieties of plant and animal life in an illustrated book. While you do so, take notes regarding the ways that certain species seem to be related. Prepare a list of the species you have observed and classify them into groups, explaining the common physical characteristics they share. Then, with the help of a teacher, check your classification system and see if it agrees with the established system of classification in use by scientists today.

(6) One of the most severe problems affecting premodern life was the lack of any consistent public health policy, a fact that resulted in the spread of disease. Consider the ways that public health is applied in your community through methods such as sanitation, education, treatment, and preventive medicine. List these different methods, and think of ways that they could be improved—as well as the reasons why some people might oppose these improvements.

FOR MORE INFORMATION

Books

Anderson, Margaret J., and Karen F. Stephenson. *Scientists of the Ancient World*. Springfield, NJ: Enslow Publishers, 1999.

Attenborough, David. *Discovering Life on Earth: A Natural History*. Boston: Little, Brown, 1981.

Beshore, George. *Science in Ancient China*. New York: F. Watts, 1998.

Cannarella, Deborah, and Jane Fournier. *Medicine*. Vero Beach, FL: Rourke Press, 1999.

Corzine, Phyllis. *The Black Death*. San Diego, CA: Lucent Books, 1997.

Garza, Hedda. *Women in Medicine*. New York: F. Watts, 1994.

Gates, Phil, and Ghislaine Lawrence. *The History News: Medicine*. New York: Scholastic, 1997.

Gay, Kathlyn. *Science in Ancient Greece*. New York: F. Watts, 1998.

Gutnik, Martin J. *The Science of Classification: Finding Order Among Living and Nonliving Objects*. New York: F. Watts, 1980.

Hooper, Mary, and Greg Gormley. *Bodies for Sale*. New York: F. Watts, 1999.

Hunter, Shaun. *Leaders in Medicine*. New York: Crabtree Publishing, 1999.

"Inventors and Inventions." In *Medicine and Health*. Vol. 8. Danbury, CT: Grolier Educational, 2000.

January, Brendan. *Science in Colonial America*. New York: F. Watts, 1999.

Jennings, Gael, and Roland Harvey. *Sick!: Bloody Moments in the History of Medicine*. New York: F. Watts, 2000.

Mulcahy, Robert. *Medical Technology: Inventing the Instruments*. Minneapolis, MN: Oliver Press, 1997.

Nardo, Don. *Scientists of Ancient Greece*. San Diego, CA: Lucent Books, 1999.

Royston, Angela. *100 Greatest Medical Discoveries*. Danbury, CT: Grolier Educational, 1997.

Spangenburg, Ray, and Diane Moser. *The History of Science from the Ancient Greeks to the Scientific Revolution*. New York: Facts on File, 1993.

Stewart, Melissa. *Science in Ancient India*. New York: F. Watts, 1999.

Stille, Darlene R. *Extraordinary Women of Medicine*. New York: Children's Press, 1997.

Storring, Rod. *A Doctor's Life: A Visual History of Doctors and Nurses Through the Ages*. New York: Dutton Children's Books, 1998.

Ward, Brian R. *The Story of Medicine: Medicine Around the World and Across the Ages*. New York: Lorenz Books, 2000.

Woods, Geraldine. *Science in Ancient Egypt*. New York: Franklin Watts, 1998.

Life Science

FOR MORE INFORMATION

Woods, Michael, and Mary B. Woods. *Ancient Medicine: From Sorcery to Surgery.* Minneapolis, MN: Runestone Press, 2000.

Web sites

"Aristotle—Overview." *Internet Encyclopedia of Philosophy.* http://www.utm.edu/research/iep/a/aristotl.htm (accessed on February 12, 2002).

"The Black Death: Bubonic Plague." *Brigham Young University.* http://www.byu.edu/ipt/projects/middleages/LifeTimes/Plague.html (accessed on February 12, 2002).

"Biotech Chronicles." *Access Excellence.* http://www.accessexcellence.com/AB/BC/ (accessed on February 12, 2002).

"A Brief History of Marine Biology and Oceanography." *Marine and Environmental Education and Research.* http://www.meer.org/mbhist.htm (accessed on February 14, 2002).

Carr, Ian. "Some Obstetrical History." *University of Manitoba.* http://www.umanitoba.ca/outreach/manitoba_womens_health/hist1.htm (accessed on February 12, 2002).

"Chronology of Horticulture." http://www.hcs.ohio-state.edu/hcs/TMI/HCS210/HortOrigins/OutlineHistory.html (accessed on February 12, 2002).

Cockren, A. "Paracelsus." *Alchemy Lab.* http://www.alchemylab.com/paracelsus.htm (accessed on February 12, 2002).

"The Four Humors." *About.com.* http://ancienthistory.about.com/library/weekly/aa020299.htm (accessed on February 12, 2002).

"Galen of Pergamon." http://www.chorlton.co.uk/Yasmin/. (accessed on February 12, 2002).

"History of Acupuncture in China." http://www.acupuncturecare.com/acupunct.htm (accessed on February 12, 2002).

"History of Biology." http://www.ucmp.berkeley.edu/help/topic/history.html (accessed on February 12, 2002).

"History of the Health Sciences World Wide Web Links." http://www.mla-hhss.org/histlink.htm (accessed on February 12, 2002).

"History of Science." http://biology.clc.uc.edu/courses/bio104/hist_sci.htm (accessed on February 12, 2002).

"History of Traditional Indian Medicine." *Karolinska Institutet.* http://www.mic.ki.se/India.html (accessed on February 12, 2002).

"The Hospital of San Luca in Lucca." http://www.sns.it/html/Groups/Archeo/S.Luca/Hosp.html (accessed on February 14, 2002).

The Museum of Questionable Medical Devices. http://www.mtn.org/~quack/ (accessed on February 14, 2002).

"Muslim Scientists, Mathematics and Astronomers." http://users.erols.com/zenithco/index.html (accessed on July 23, 2001).

"Plague and Public Health in Renaissance Europe." http://jefferson.village.virginia.edu/osheim/intro.html (accessed on February 14, 2002).

"Science Fiction Timeline." *The Ultimate Science Fiction Web Guide.* http://www.magicdragon.com/UltimateSF/timeline.html (accessed on February 14, 2002).

Life Science

FOR MORE
INFORMATION

Index

Italic type indicates volume number; **boldface** indicates main entries and their page numbers; (ill.) indicates photos and illustrations.

A

Abacus 2:164, 165 (ill.), 185 (ill.)
Abbas, Haly 1:132
Abelard, Peter 1:82, 2:335
Abu Ya'qub Yusuf 1:111
Abu'l-Wefa 2:152, 226
Académie Royale des Sciences 2:262
Acid-base indicators 2:374
Acoustics 2:393
Actuarial science 2:219
Acupuncture 1:19, 21 (ill.), 23 (ill.)
    China 1:20–22
Adanson, Michel 1:85
Adelard of Bath 2:330
Aeolipile 3:564
Aerodynamics 3:535
African American scientists and inventors
    Banneker, Benjamin 3:559
    Derham, James 1:129
Agamede 1:72

Ages of man (stone age, bronze age, iron age) 3:422–427
Agnesi, Maria Gaëtana 2:254
Agnodice 1:72
Agricola, Georgius 2:365, 390
Agriculture 3:430–439, 431 (ill.)
    crop rotation 3:491
    Mesopotamia 3:412
    seed drill 3:574
    tools and techniques 3:437–439
    water management 3:437
Ahmes 2:162, 254
Air pump 3:563
Airfoil 3:535
Airplanes 3:524, 535
Airships 3:535
Al-Battani 2:328, 391
Al-Bitruji 2:326
Al-Hajjaj 2:259
Al-Hawi 1:39
Al-jabr 2:152, 247
Al-Jazari 3:566
Al-Juzjani, Abu Ubayd 2:326
Al-Kashi 2:260
Al-Khwarizmi 2:152, 188, 206, **246–248**, 247 (ill.)
    algebra 2:210
    trigonometry 2:328

Index

Al-Kindi 2:323, 370, 398
Al-Ma'mun 2:320
Al-Majisti. See Almagest
Al-Maliki 1:132
Al-Qanun fi al Tibb 1:113
Al-Shirazi, Qutb al-Din 2:324
Al-Sufi, Abd al-Rahman 2:401
Al-Tusi, Nasir Ad-Din 2:327, 402
Albategnius. *See* Al-Battani
Albert of Saxony 2:390
Albertus Magnus 1:8, 82, 126, 2:334
   optics 2:357
   rainbows 2:329
Alchemy 1:10 (ill.), 2:281, 282 (ill.), 339–342
   China 3:498
   False Geber 2:340, 401
   Geber 2:340, 401
   Medieval Europe 2:331–342
   Rhazes 2:400
Alcuin 2:333
Alexander the Great 3:570
Algebra 2:151, 210–213
   Agnesi, Maria Gaëtana 2:254
   Al-Khwarizmi 2:210
   Aryabhatiya 2:235
   Brahmagupta 2:212
   Chi'in Chiu-shao 2:166
   China 2:268
   cubic equations 2:212
   notation 2:268
Algebraic equations 2:237
   Takakazu Seki Kowa 2:266
Algorithm 2:246
Alhazen 2:280, 323, 357, **369–372**
Ali ibn Abbas al-Majusi, 132
Alkahest 2:340
Almagest 2:152, 226, 259, 316, 325, 388
   revisions 2:403
   translations 2:330

Alphabet 3:414, 444
Alphabetic notation 2:182
Alternative Medicine 1:19–28, 48
Amputation 1:107
Anacharsis of Scythia 3:563
Analytic geometry. *See* Coordinate geometry
Anatomy 1:55
   Da Vinci, Leonardo 1:134, 3:550
   Dei Liucci, Mondino 1:135
   Guy, de Chauliac 1:131
   microscopic 1:134
   Swammerdam, Jan 1:138
   textbooks 1:9
   Vesalius, Andreas 1:124
   See also Dissection
Anaximander 2:299
Anaximenes 2:299
Anderson, Andrew 3:556
Andes, development of cities in 3:454
Angle, trisection of 2:201
Angles, curves and surfaces 2:223–235
   See also Trigonometry
Angstrom units 2:177
Animal circulatory system 1:132
Animal electricity, theory of 1:91, 92, 130
Animal magnetism, theory of 1:137
Animalcules 1:119
Animals, domestication of 3:435–437
Anne of Bohemia 3:559
Anothomia 1:135
Antiphon 2:200
Apollonius of Perga 2:203, 254
Apotelesmatica 2:389
Apothecaries 1:44, 48
Appian Way 3:469, 559

Index

Appius Claudius Caecus 3:559
Applied mathematics 2:172
Aqua regia 2:340
Aqueducts 3:416, 462
Aquinas, Thomas 2:281, 311 (ill.), 313, 334
Ar-Razi. *See* Rhazes
Arabic numbers. *See* Hindu-Arabic numeral system
Archimedes 2:201, 254, 3:415 (ill.), 471, 525 (ill.), 536–538, 537 (ill.)
   buoyancy 3:524–526
   inventions 3:416
   levers and planes 3:471
   pi 3:536
   water screw 3:472, 473 (ill.)
Archimedes' screw 3:472, 473 (ill.), 536
Architecture
   Imhotep 3:547
   Ponte Vecchio 3:562
   St. Peter's Basilica 3:567
   Vitruvius 3:574
Archytas of Tarentum 2:254
Aristarchus of Samos 2:224, 226, 314, 315, 374, 390
   heliocentric theory 2:278
Aristotle 1:2, 5, 6, 8, 12, 38, 41, 79–81, 84, 85, 98, **108–110**, 109 (ill.), 110 (ill.), 138, 2:150, 175, 296, 393
   Buridan, Jean 2:346
   four element theory of matter 2:278
   influence on science 2:308–320
   logic 2:177–178
Arkwright, Richard 3:559
Armati, Salvino degli 3:572
Arnau de Villanova 2:342
Ars magna 3:558

Aryabhata 2:149, 170, **235–236**, 280, 294
Aryabhatiya 2:170, 235, 294
Asclepiades of Bithynia 1:32, 126, 2:307
Asclepias 1:29
Astadhyayi 2:265
Astangahrdaya Samhita 1:24
Astrolabe 2:327
Astrology 1:44, 46, 2:287, 338
   Egypt 2:288
   India 2:293
   medicine 2:341
   Medieval Europe 2:331–342
   Mesopotamia 2:288
   See also Astronomy
Astronauts 2:353 (ill.)
Astronomical clock 3:418, 506, 551, 570
Astronomy 2:**285–297**
   Al-Battani 2:391
   Al-Tusi, Nasir Ad-Din 2:402
   Babylonia 2:275
   Brahe, Tycho 2:391
   China 2:291
   Enlightenment 2:367
   Herschel, William 2:396
   India 2:294
   mathematics 2:174
   observatories 2:326, 402
   Regiomontanus 2:403
   Shih Shen 2:400
   Spain 2:326
   Su-sung 3:551
   tables 2:327
   timekeeping 3:497
   tools 2:327
Aswan High Dam 2:370, 371 (ill.)
Athens 3:454
Atomic structure 2:305 (ill.)
Atomic theory 1:126, 2:304–308
   Boyle, Robert 2:373

Index

Dalton, John 2:307, 356
Democritus 2:277, 308, 394
India 2:293
Atomism. *See* Atomic theory
The Audidi 1:43
Aurora borealis 3:544
Automobile 3:573
Avenzoar 1:126
Averroës 1:6, 36, 39, 41, **110–112**, 110 (ill.), 2:280, 326, 334, 390
Avicenna 1:2, 6, 36, 41–42, **112–113**, 113 (ill.), 122, 2:280, 324, 346, 390, 398
Axle and wheel 3:475–477, 476 (ill.)
Ayurveda 1:24

B

Babylonia
    astronomy 2:275
    calendar 2:296, 3:503
    water management 3:460
    zero 2:189
Bacon, Roger 2:281, 329, 342, 357, 3:**538–540**, 539 (ill.)
    Grosseteste, Robert 2:395
    optics 2:331
Ballistics 2:254
Balloons 3:523, 532, 533 (ill.)
Bamboo 3:435, 435 (ill.), 444
Bandages 1:19
Banneker, Benjamin 3:559
Baptism of Christ 3:550
Baptista van Helmont, Jan 1:94
Barber-surgeons 1:47
Barometer 2:402, 3:564
Base-5 system 2:180
Battani. *See* Al-Battani.
Batteries 2:364
The Battle of Ten Naked Men 1:66
Bauer, Georg. *See* Agricola, Georgius

Beaulieu, Jacques de 1:53
Becher, Johann 2:359, 359 (ill.)
Becket, Thomas 1:50, 51 (ill.)
Bedlam 1:8,9 (ill.)
Beg, Ulugh 2:266
Bent Pyramid 3:449
Berengario da Carpi, Giacomo 1:68
Berkeley, George 2:232
Bernoulli, Daniel 2:234, 242, 356, 3:524, 532, 535, **540–542**, 541 (ill.)
    fluid mechanics 2:285
Bernoulli, Jakob 2:228, 234, 241, 255, 3:541
Bernoulli, Johann 2:228, 234, 241, 257, 3:541
Bernoulli's principle 2:257, 3:532, 535, 536, 540, 567
Bestiaries 1:80, 84
Bhaskara 2:170, 222, 255
Biblia naturae 1:138
Bichat, Marie-François-Xavier 1:136
Bifocal eyeglasses 3:543
Big Bang theory 2:394
Bile 1:31
Binary numbers 2:181
Bingen, Hildegard von 1:73, 82, 133
Binomial nomenclature 1:13, 84, 85
Binomial theorem 2:386
Biology and Classification 1:**79–85**
Bireme 3:487
Bitruji. *See* Al-Bitruji
Black Death 1:9, 56, 57–59, 58 (ill.)
Black, Joseph 2:361, 383, 3:556
Blane, Gilbert 1:103
Blast furnace 3:562

Index

Block printing 3:417, 442, 494, 495, 514
Block-and-tackle system 3:475
Blood circulation 1:34
Bloodletting 1:32, 47 (ill.)
Boat and shipbuilding 3:482–490
    anchors 3:563
Boerhaave, Hermann 1:106, 126
Boethius 2:255
Bolyai, Janos 2:207
Bonrepos, Pierre-Paul Riquet de 3:469, 571
Book of Furnaces 2:401
Book of Optics 2:323
The Book of Simple Medicine 1:73, 133
Boomerang 3:536
The Boston News-Letter 3:522 (ill.)
Botanical gardens 1:83
Botany 1:81
    illustrations 1:67, 68
Boulton, Matthew 3:529, 557
Bourgeois, Louis 1:78, 127
Bourne, William 3:487
Boyle, Robert 1:97, 2:281, 306, 342, 344, **372–374**, 373 (ill.), 3:527 (ill.)
Boyle's law 2:373
Bradwardine, Thomas 2:346, 391
Brahamasphuta Siddhanta 2:192
Brahe, Tycho 2:391
    Kepler, Johannes 2:381
    scientific revolution 2:344
Brahmagupta 2:192, 212, 256, 294
Brahmans 1:28, 90 (ill.)
Bramah, Joseph 3:560
Bramante, Donato 3:567
Brand, Hennig 2:306
Brass 3:426
Breech births 1:74
Bridges 3:562
Briggs, Henry 2:214, 267

Bronze 3:426
Bronze Age 3:424
Brunfels, Otto 1:68
Bruno, Giodano 2:283, 349
Bubonic plague. *See* Black Death
Buddhism and Hindusim 1:22–25
Buffon, Georges Louis Leclerc de 1:85, 2:366, 392
Buoyancy 3:524–526, 569
Buridan, Jean 2:345, 346, 392, 398
Byzantine Empire 1:45

C

Cadavers, dissection of. *See* Dissection
Caduceus 1:29, 30 (ill.)
Caecus, Appius Claudius 3:559
Calculators
    China 2:166
    mechanical 2:250, 251 (ill.), 267
Calculus 2:156, 230–235
    Newton, Issac 2:385
    Newton-Liebniz controversy 2:232, 263
Calendar
    Babylonia 2:296, 3:503
    India 2:293
    Mayan 2:169
    Middle East 2:325
    reform 2:399, 3:504, 510, 560
Calligraphy 2:328 (ill.)
Callinicus of Heliopolis 3:560
Callippus 2:311
Caloric theory 2:360–2:362
Calorimeter 2:384, 397
Calvin, John 3:519
Campanus of Novara 2:330
Canal du Midi 3:466, 468, 571
Canals 3:463
    Bonrepos, Pierre-Paul Riquet de 3:571

Index

China 3:494
Cancer 1:13, 100, 103, 136
Canon of Medicine 1:40, 130
Canterbury Tales 1:50
Carbon dioxide 2:397
Carbonation 2:360
Cardano, Girolamo 2:218, 260
Cartesian coordinate system 2:225, 228, 237
Cassiodorus, Flavius Magnus Aurelius 3:560
Caste system, India 1:28
Catapult 3:538
Cataracts 1:106
Cavendish, Henry 2:354, 384
Cavities 1:12
Cellier, Elizabeth 1:78
Celsius, Anders 2:285, 361, 392
Celsus 1:127
Centripetal force 2:380
Cerebri anatomi 1:139
Cesalpino, Andrea 1:84
Cesarean section 1:71, 74, 106
Ceulen, Ludolph van 2:260
Ch'iao Wei-yo 3:561
Ch'in Chiu-Shao 2:166, 256
Ch'in dynasty 3:492, 500
Ch'in Shih Huang Ti 3:464, 500
Chamberlen the Elder, Peter 1:75, 127
Chang Heng 3:560
Charaka Samhita 1:24
Charlemagne 2:333, 335 (ill.)
Chaucer, Geoffrey 1:50
Chavín de Huántar 3:454
Chemical nomenclature 2:384
Chemical reactions
    Black, Joseph 2:397
Chemical reactions, conservation of matter in 2:383
Chemical theory 2:383
Chemistry
    Boyle, Robert 2:373
    Lavoisier, Antoine-Laurent 2:284
Chen Chuan 1:127
Chen Zhuo 2:401
Chi. *See* Qi
Chichen Itza 2:193 (ill.)
Chimney sweeps 1:103–104
China 3:411
    astronomy 2:291
    boat and shipbuilding 3:496
    calculators 2:166
    canals 3:463, 494
    combinatorics 2:220
    Grand Canal 3:561
    Great Wall 3:464, 492, 500–501
    gunpowder 3:419
    ink 3:492
    inventions 3:417
    magnetic compass 3:419
    paper and printing 3:417, 494–495, 514
    physical sciences 2:280
    pi 2:148
    rod numerals 2:187
    seismology 2:365
    technology and invention 3:490–502
    timekeeping 3:418
    wheelbarrow 3:476, 493
    writing 3:443
China, mathematics in 2:163–168
Chinchon, Ana (de Osorio), Countess of 1:48
Chinese medicine 1:134
Chirurgia 1:56
Chirurgia magna 1:57, 122, 131
Chladni, Ernst 2:393
Choleric personality 1:33
Chou Dynasty 3:492
Chu Hsi 2:394

Index

Chu Shih-Chieh 1:256
Chu tsi shu chieh yao 2:394
Circle 2:259
Circle, squaring of 2:200
The Circles of Proportion and the Horizontal Instrument 2:263
Circulatory system 1:96–97, 116
Cities and structures, early development of 2:396, 3:447–459
    infrastructure 3:459–470
Civil engineering 3:572
Civil service 3:492
Civilization, development in the fertile crescent 3:411
Classification, Biology and 1:79–85
Clavis mathematicae 2:263
Clepsydra 3:478, 508 (ill.), 561
Clinical drug trials 1:41
Cloaca Maxima 3:462
Clocks 3:498, 506
    See also Timekeeping
Colinet, Marie 1:73
The Collection of the Essence of the Eight Limbs of Ayurveda 1:24
Color spectrum 2:386
Colossus of Rhodes 3:458
Columbus, Christopher 1:86, 2:389
Columns 3:456
Combinatorics 2:220–222
    Fibonacci sequence 2:246
    Königsberg bridge problem 2:222
Combustion 2:360
    oxygen 2:382, 384
    technology 3:412
Comets 2:291, 354
Commentariolus 2:376
Communia mathematica 3:540
Communia naturalium 3:540

Compass 3:399, 417, 419, 496–497, 567
Compendium philosophiae 3:540
Computer 2:155, 3:558
Computer-aided design (CAD) 2:230
Condorcet Paradox 2:256
Condorcet, Marie-Jean-Antoine-Nicholas de, Marquis de Condorcet 2:256
Confucius 1:21
Conic sections 2:203, 254
Conservation of momentum 2:378, 379
The Consolation of Philosophy 2:255
Constantine the African 1:127
Contagion, theories of 1:59
Coordinate geometry 2:244, 263
Coordinate systems 2:228
Copernicus, Nicolaus 2:236, 283, 313, 347–349, **375–376**, 375 (ill.)
    astrology 2:338
    heliocentric theory 2:390
    scientific revolution 2:344
Copper 3:426
Copyright laws 3:520
Coranto 3:521
Corn 3:435
Corpuscular theory of light 2:344, 356, 386
Cortés, Hernan 3:425 (ill.)
Cosmology 2:287
    Al-Tusi, Nasi Ad-Din 2:402
    Aristotle 2:311
    Greece 2:278
    Mesopotamia 2:291
    Ptolemy 2:278
Cotton gin 3:574
coulomb 2:394

Index

Coulomb, Charles-Augustin 2:364, 394
Coulomb's Law 2:394
Counting rods 1:166, 2:262
Cowpox 1:27
Cristofori, Bartolemeo di Francesco 3:561
Crop rotation 3:491
Crusades 1:6, 7 (ill.), 50
Cryfts, Nicholas. See Nicholas of Cusa
Cryptography 2:209
Ctesibius of Alexandria 3:478, 508, 561
Cube, doubling of 2:202
Cube roots 2:163, 164
Cubic equations 2:152, 212
    Omar Khayyam 2:154
    Tartaglia-Cardano feud 2:260
Cuneiform 2:148 (ill.), 149 (ill.), 159 (ill.), 182, 3:442 (ill.)
Curie, Pierre 3:497
Curie point 3:497
Curved surfaces, representation of 2:229
Curves. See Angles, Curves and Surfaces
Cuvier, Georges 1:85

D

D'Alembert, Jean le Rond 2:234, 258
D'Oresme, Nicole 2:231
Da Vinci, Leonardo 1:64, 66, 134, 2:201, 365, 3:429, **549–551**, 550 (ill.)
    pendulum 3:509
    submarines 3:486
Dalton, John 2:277, 307, 356
Darby, Abraham 3:562
Dark Ages 1:45
Darwin, Charles 1:13, 85, 92

David 1:65 (ill.), 3:567
Davy, Humphrey 2:363
Day, Stephen 3:517
De animalibus 1:82
De architectura 3:574
De Causis Plantarum 1:81, 138
De generatione animale 1:116
De historia stirpium 1:68, 84
De humani corporis fabrica 1:9, 69, 124, 125
De materia medica 1:40 (ill.), 128
De medicina 1:127
De Morgan, Augustus 2:201
De motu 2:377
De re metallica 2:390
De revolutionibus orbium coelestium 2:349, 376
De situ orbis 2:399
De vegetabilibus de plantis 1:82
Decimal fractions 2:164, 260
Decimal system 2:158, 179
Dei Liucci, Mondino 1:135
Delian problem. See Doubling the cube
Demetrius of Phaleron 2:198
Demiurge concept 2:310
Democritus 2:277, 304, 394
Dentistry
    cavities 1:12
    dental drill 1:129
    dentures 1:108, 129
    Eighteenth century 1:106, 108
Derham, James 1:129
Desargues, Girard 2:250, 258
Descartes, Réne 1:11, 14, 52, 93, 94 (ill.), 96, 97, 128, 2:**236–238**, 2:236 (ill.)
    algebra 2:213
    algebraic equations 2:237
    cartesian coordinate system 2:228, 237
    mathematical notation 2:194

scientific revolution 2:344
    trisecting an angle 2:202
Diabetes mellitus 1:127
Dialogue Concerning Two Chief World Systems 2:349, 378
Diamond, Jared 3:437
Dickens, Charles 1:104
Diderot, Denis 2:258
Dietrich, von Freiberg 2:206, 325
Differential calculus 2:230, 243, 249
Diophantus of Alexandria 2:152, 211, 244, 258
Dioscorides, Pedanius 1:82, 128
Discobolus 1:66 (ill.)
Discourse on Method 2:237
Disease
    prevention of 1:100
    spread of 1:88
Dissection 1:8, 29, 52 (ill.), 56, 61
    Frederick II 1:131
    Galen 1:34
    grave robbing 1:69
    India 1:27
    prohibitions against 1:66
Distillation 2:340
Diving bell 3:486 (ill.)
Division symbol 2:194
DNA (deoxyribonucleic acid) 1:99
Dodecahedron 2:319
Dog, domestication of 3:435
Dolphin 1:4 (ill.)
Domestication 3:433–437
Double boiler 2:339, 399
Doubling the cube 2:202–203
    Eratosthenes 2:239
    Hippocrates of Chios 2:259
Drebbel, Cornelis Jacobszoon 3:486, 562
Dufay, Charles-François 2:364
Dürer, Albrecht 1:66–68

E

e 2:215
Earth
    age of 2:366, 369
    circumference of 2:238, 294, 315
    Erastosthenes 2:152, 225, 233, 389
    Ptolemy 2:152, 233, 389
    magnetic field 2:364
    rotation of 2:294
    sedimentary layers 2:366
    temperature zones 2:399
Earth, center of the universe. *See* Geocentric theory
Earth sciences. *See* Geology
Earthquakes, detection of 3:560
Ecole Polytechnique 2:230
Edison, Thomas 3:429
Edward I 3:562
Egypt
    astrology 2:288
    boat and shipbuilding 3:483, 485
    canals 3:463
    cities and pyramids 3:448–451, 548
    hieroglyphics 3:442, 443
    Imhotep 3:548
    mathematics 2:162, 180
    mummification and medicine 1:14–19
    trigonometry 2:224
    writing 3:440
Eighteenth century
    dentistry 1:106, 108
    Enlightenment 1:354–369
    health care 1:99–108
    hospitals 1:106
    life sciences 1:12
    surgery 1:106
Einkommende Zeitung 3:520

Index

Einstein, Albert 2:207 (ill.)
Elastic collisions 2:378, 379
Electric charges 3:545
Electric current 2:364
Electrical experiments 2:364
Electromagnetic theory 2:363
Electromagnetism 2:234, 354, 363
    Coulomb, Charles-Augustin 2:394
Elements 2:306, 342, 374
    naming conventions 3:428
    Scheele, Karl Wilhelm 2:400
    table of 2:301 (ill.)
Elements of Geometry 2:203–204, 239, 259
    commentaries on 2:266
    translations 2:206, 259
Embalming. *See also* Mummification
Embodiment, theory of 1:98
Embryo 1:93, 98 (ill.)
Empedocles 1:5, 31, 117
    four-elements theory 2:304
Encyclopedia 3:493, 572
Encyclopédie 2:258
Englightenment 1:99, 2:215 (ill.), 354–369
Entomology 1:138
Enzymes 1:94
Epicurus 2:307
Epigenesis 1:14, 97–98, 116
Epistola de magnete 2:399
Epitomae medicae libri septem 1:72
Epitome of Astronomy 2:347, 403
Equals symbol 2:194, 266
Equinoxes 2:259
Erasistratus 1:128
Eratosthenes 2:152, 238–239
    doubling the cube 2:239
    Earth, circumference of 2:233, 314, 315
    Library of Alexandria 2:199
    prime numbers 2:208
Ericsson, John 3:472
Essai sur les coniques 2:250
Essay on the Application of Analysis to the Probability of Majority Decisions 2:256
Euclid 2:150, 203, **239–241**, 240 (ill.), 346
    Elements of Geometry 2:203, 205 (ill.)
    Library of Alexandria 2:199
    logic 2:204
    optics 2:323
    parallel postulate 2:240
Euclidean geometry 2:196, 204
    unsolved problems in 2:150
Eudoxus of Cnidus 2:310, 315, 326
Euler, Leonhard 2:156, 194, 201, **241–243**, 241 (ill.)
    Bacon, Roger 3:541
    Fermat's last theorem 2:244
    Könisberg Bridge Problem 2:222, 241
    three-dimensional coordinate systems 2:228
    variational calculus 2:234
Europe, life sciences in 1:54–61
Eustachian tube 1:128
Eustachio, Bartolommeo 1:11, 128
Exercitatio anatomica de motu cordis sanguinis in animalius 1:95, 96, 116
Exercitationes quaedam mathematicae 3:541
Experimental History of Colors 2:374
Exploration of new lands, impact on the sciences 1:85–92
Exponents 2:155
Extromission theory 2:323, 370
Eyeglasses 3:543, 572

Index

F

Fabiola 1:35, 72, 76
Fabrici, Girolamo 1:73
Fahrenheit, Daniel 2:285, 361, 395
Fallopian tubes 1:75, 129
Falloppio, Gabriele 1:11, 75, 129
False Geber 2:340, 401
Farisi, Kamal al-Din 2:324
Farming. *See* Agriculture
The Fates of Human Societies 3:437
Fauchard, Pierre 1:12, 108, 129
Female reproductive system 1:131
    See also Women, medical discovery of
Ferari, Ludovico 2:261
Fermat, Pierre de 2:156 (ill.), 209, 218, 236, **243–245**, 243 (ill.)
    algebra 2:213
    Cartesian coordinate system 2:228, 244
    Euler, Leonhard 2:243
Fermat's Last Theorem 2:244, 264
Ferments 1:94
Ferro, Sciopione dal 2:261
Fertile Crescent 3:412, 413 (ill.), 431
Fibonacci sequence 2:245, 246
Fibonacci, Leonardo 2:8, 154, 188, 222, **245–246**, 245 (ill.)
Fidelis, Cassandra 1:73
Fifth postulate. *See* Parallel postulate
Fior, Antonio 2:261
Fire, phlogistan theory of 2:358, 359
Fire, Metal, and Tools 3:**421–430**
First law of motion 2:346
    Buridan, Jean 2:392
    Philoponus, Johannes 2:398
First law of planetary motion 2:352
Five-elements theory 2:403
Flint blades 3:422 (ill.)
Flora Lapponica 1:120
Fluid dynamics. *See* Hydrodynamics
Fluid mechanics 2:234
Fluids, Pressure, and Heat 3:**523–536**
Flying shuttleloom 3:531
Fontana, Niccolò. *See* Tartaglia
Foods, introduction from the Americas 1:86
Forceps 1:71, 75, 77 (ill.)
Forty-Two Line Bible. *See* Gutenberg Bible
Fossils 1:136, 2:365, 394
Founding fathers of science
    father of acoustics and meteoritics. *See* Chladni
    father of algebra. *See* Al-Khwarizmi
    father of biology. *See* Aristotle
    father of botany. *See* Theophrastus
    father of chemistry. *See* Boyle
    father of chemistry. *See* Lavoisier
    father of experimental science. *See* Archimedes,
    father of histology. *See* Malpighi
    father of medicine. *See* Hippocrates
    father of mineralogy. *See* Agricola
    father of modern geology. *See* Hutton
    father of modern military surgery. *See* Paré
    father of modern philosophy. *See* Descartes
    father of modern surgery. *See* Hunter

father of philosophy and the physical sciences. *See* Thales of Miletus
father of physiology. *See* Erasistratus
father of physiology. *See* Galen
father of plant physiology. *See* Hales
father of scientific dentistry. *See* Fauchard
father of trigonometry. *See* Hipparchus
fathers of calculus. *See* Liebniz and Newton
fathers of entomology. *See* Réaumur and Swammerdam
Foundling Hospital of London 1:105
Four-element theory 2:277, 310, 393
Fracastoro, Girolamo 1:59, 60 (ill.), 62, 130
Fractional exponents 2:263
Frame of reference 2:347
Frank, Johann Peter 1:104, 130
Franklin stove 3:543
Franklin, Benjamin 2:285, 364, **3:542–545**, 543 (ill.)
Frederick I Barbarossa 2:333
Frederick II 1:8, 131
Freiberg, Dietrich von 2:206
French revolution 2:351
    Lavoisier, Antoine-Laurent 2:384
Frère Jacques 1:53
Fuchs, Leonhard 1:68, 84
Fust, Johann 3:546

G

G 2:353
Gaddi, Taddeo 3:562
Galen 1:5, 6, 29, 32, 34–35, 38–42, 45, 46, 66, 70, 96, **113–115**, 114 (ill.), 116, 122, 124, 125, 128, 132, 2:360
Galilei, Galileo. *See* Galileo
Galileo 1:88, 2:313, 356, **376–378**, 377 (ill.)
    astrology 2:338
    gravity 2:350
    heliocentric theory 2:283, 376
    pendulum 3:509
    scientific method 2:350, 377
    scientific revolution 2:344
    telescope 3:349, 566
Galindo, Beatrix 1:73
Galleys 3:486
Galvani, Luigi 1:91, 130
Gambling 2:217, 218 (ill.)
Gan De 2:401
Gases 2:373
Gassendi, Pierre 2:356
gauss 2:258
Gauss, Carl Friedrich 2:201, 207, 258
Gearwheel 3:478
Geber 2:280, 401
    alchemy 2:340
Gellibrand, Henry 2:364
Geminus 2:259
Geocentric theory 2:278, 280, 300
    Aristotle 2:393
    Ptolemy 2:375
    See also Heliocentric theory
Geodesy 2:225
Geography 2:233
Geography 2:233, 389
Geology 2:284, 365
    Albert of Saxony 2:390
    Hutton, James 2:396
    Lehmann, Johann Gottlob 2:398
    Renaissance 2:365

stratigraphy 2:398
uniformitarianism 2:398
Geometry 2:227–230
    Apollonius of Perga 2:254
    Greece 2:199
    Li Yeh 2:166
    military use 2:250
    projective 2:230
    unsolved problems in 2:195–207
Gerard of Cremona 1:130, 2:206, 330
Germ theory 1:59
Germain, Sophie 2:264
Gerson, Levi ben 2:221
Gesner, Conrad 1:84, 130
Gilbert, William 2:363
Glassmaking 3:568
Gold 3:425
Graaf, Reinier de 1:77, 131
Grand Canal 3:494, 561
Graph theory 2:156, 223, 241
Grave robbing 1:69
Gravitational constant 2:353
Gravitational theory 2:386, 387
Gravity 2:279 (ill.), 352
    Galileo 2:283, 350
    India 2:293
    Newton, Isaac 2:262
Gravity clock 3:506
    See also Clepsydra, Hourglasses, Timekeeping
The Great Piece of Turf 1:67
Great Pyramid of Cheops 3:458, 549
Great Wall of China 3:464, 492, 500–501
Greater Magellanic Cloud 2:401
Greece
    cosmology 2:278
    development of cities in 3:454–459
    geometry 2:199
    influence on medicine 1:29–36
    theories of matter 2:297–308
    trigonometry 2:151
Greek alphabet 2:183 (ill.)
Greek fire 3:560
Greenwich mean time (GMT) 2:367
Gregorian calendar 3:505–506
Grosse Wund Artzney 1:122
Grosseteste, Robert 2:395, 539
Guericke, Otton von 3:563
Guillotin, Ignace 1:104
Gunpowder 3:417
    Bacon, Roger 3:540
    China 3:419, 498
Gutenberg Bible 3:515, 517 (ill.), 546
Gutenberg, Johannes 1:52, 62, 3:420, 440, 495, 512, 515–517, **545–546**, 546 (ill.), 570
Guy, de Chauliac 1:57, 131
Gynaecology 1:71
Gynecology and obstetrics 1:71–75
    midwives 1:74
    Paul of Aegina 1:135
    Rhazes 1:124

H

Hadrian's Wall 3:501
Hajjaj. See Al-Hajjaj
Hales, Stephen 1:91, 132, 2:397
Halley, Edmond 2:285, 354, 368, 387, 395
    diving bell 3:487
Halley's comet 2:367 (ill.), 368, 395, 403
Haly Abbas 1:132
Hammurabi 3:504
Han Dynasty 3:492
Hanging Gardens of Babylon 3:461 (ill.)

Index

Harappa 3:451
Harrison, John 3:511, 565
Harvey, William 1:12, 34, 41, 95, 96, **115–116**, 116 (ill.), 132, 135
Hatubi sanpo 2:266
Health care in the Eighteenth Century 1:99–108
Heat 2:358
Heat capacity 2:361
Heat transfer 2:362
Heat, Fluids, and Pressure 3:523–536
Heliocentric theory 2:300, 314, 347, 348 (ill.), 375
    Aristarchus of Samos 2:390
    Copernicus, Nicolaus 2:374
    Nicholas of Cusa 2:399
Hemp 3:492
Henlein, Peter 3:509, 564
Heracleitus 2:276, 302
Herbal medicines 1:19 (ill.)
Herbalists 1:44, 48
Herbals 1:81 (ill.), 128, 133
Herbarum vivae eicones 1:84
Hero of Alexandria 3:472, 478, 527, 564
Herschel, Alexander 2:368
Herschel, Caroline 2:368, 396
Herschel, William 2:285, 354, 368, 396
Hieratic characters 2:162
Hieroglyphics 2:182, 3:442, 443 (ill.)
Hindenburg 3:535
Hindu-Arabic numeral system 2:164, 168, 184 (ill.), **186–195**, 188 (ill.), 294
    Europe 2:154
    Fibonacci, Leonardo 2:245
    zero symbol 2:149
Hinduism and Buddhism 1:22–25

Hipparchus 2:151, **259**, 316
Hippasus 2:253
Hippocrates 1:2, 4–6, 29–32, 38, 39, 45, 46, 90, 113, 115, **117–118**, 117 (ill.), 128, 2:175
Hippocrates of Chios 2:201, 259
Hippocratic corpus 1:30
Hippocratic oath 1:4, 30, 31 (ill.), 117, 118, 2:175
Histoire naturelle 2:392
Historiae animalium 1:84, 130
Hohenheim, Philippus Bombast von. *See* Paracelsus
Hooke, Robert 1:88, 118, 139, 2:360, 365, 387, 396, 3:509
    gases 2:373
    gravitational theory 2:387
Hooke's law 2:396
Horologium oscillatorium 2:379
Horoscopes 2:287, 288, 336
Horse, domestication of 3:436, 438
Horsepower 3:555
Horseshoes 3:438
Hospitals 1:34 (ill.)
    early development of 1:35–36
    Eighteenth century 1:106
    India 1:28
    Middle East 1:43
Hourglasses 3:507
    See also Clocks, Timekeeping
House of Wisdom 2:320
Hugh of Lucca 1:56
Human body 1:92–99
    Vesalius, Andreas 1:124
    central nervous system 1:139
    circulatory system 1:11, 95 (ill.), 96–97, 116
    study of. *See* Anatomy
Human health in the Eighteenth Century 1:99–108
Human mind 1:89
Human tissue, theory of 1:136

Index

Humoral theory 1:5, 29, 31, 32, 33, 39, 40, 60, 90, 115, 117, 136
    Asclepiades of Bithynia 1:126
    India 1:25
Hunter, John 1:132
Hutton, James 2:366, 396
Huygens, Christiaan 2:344, 356, **378–380**, 379 (ill.), 3:527, 569
    light, speed of 2:358
    pendulum clock 3:509
Hydraulic press 2:250, 251, 3:471, 526
    Bramah, Joseph 3:560
    Pascal, Blaise 3:524
Hydraulica 3:541
Hydrochloric acid 2:400
Hydrodynamica 3:535, 541
Hydrodynamics 2:234, 356, 3:524, 535, 542
Hydrostatics 3:524, 537
Hypatia 1:72, 2:199, 259
Hyphegesis geographike 2:316
Hypnotism 1:137
Hypotenuse 2:172, 226
Hysterectomy 1:71
Hysteria 1:71

I

I Ching 2:220
Iatrochemistry 1:91
Iatromechanism 1:93
Iatrophysics 1:128
Ibn al-Haytham. *See* Alhazen
Ibn an-Nafis 1:41, 132
Ibn Bajja 2:346
Ibn Firnas 3:569
Ibn Rushd. *See* Averroës
Ibn Sina. *See* Avicenna
Ibn Tufayl 2:326
Ibn Yunus 2:328
Ice cream 3:492 (ill.)
Icosahedron 2:319
Ideograms 3:443
Ilkhanic Tables 2:402
Illuminated manuscript 3:514 (ill.)
Imhotep 3:449, **547–549**, 548 (ill.)
Impetus theory 2:346, 390
Incandescent light bulb 3:429
Inclined plane 3:416, 471–472
Incunabula 3:515
India
    astronomy 2:293–294
    combinatorics 2:221
    development of cities in 3:451–452
    physical sciences 2:280
    trigonometry 2:226
    zero 2:192
India, mathematics in 2:**168–171**
Industrial Revolution 3:421, 530–531, 534
    steam engine 3:524
    water frame 3:559
    Watt, James 3:557
    Whitney, Eli 3:574
Inertia 2:343, 378
Infrared radiation 2:396
Infrastructure 3:**459–470**
Ingenhousz, Jan 1:134
Ink 3:492
Inoculation 1:100, 101 (ill.)
    Jenner, Edward 1:134
Inquiry into Plants 1:81, 138
Inscribe 2:196
Insurance industry, statistics and probability theory in 2:219
Integral calculus 2:230
Intelligencer 3:521
Intercalation 3:503
Introduction to Mathematical Studies 2:256
Intromission theory 2:323, 370
    Al-Kindi 2:398

Index

Invention of Verity 2:401
The Investigation of Perfection 2:401
Iron Age 3:424, 426
Iron tools 3:414
Irrational numbers 2:172, 196, 240, 253

J

Jabir ibn Aflah. *See* Geber
Jabir Ibn Hayyan. *See* Geber
Jainism 2:293
Jansen, Sacharias 3:564, 566
Jazari. *See* Al-Jazari
Jenner, Edward 1:27, 102, 134
Johannes Philoponus. *See* Philoponus, Johannes
John of Arderne 1:57
John the Grammarian. *See* Philoponus, Johannes
Jones, William 2:194
Julian calendar 3:505
Juzjani, Abu Ubayd Al. *See* Al-Juzjani, Abu Ubayd

K

Kan Te 2:401
Kapha 1:25
Kashi. *See* Al-Kashi
Kay, John 3:559
Kepler, Johannes 2:344, 350–352, 380–382
    astrology 2:338
    Brahe, Tycho 2:391
    laws of planetary motion 2:351
    vision 2:357
Khwarizmi. *See* Al-Khwarizmi
Kindi. *See* Al-Kindi
Kinematics 2:231, 309, 343
Kinetic energy 2:346, 390
Kitab al-hawi 1:124
Kitab al-jabr wa al-muqabala 2:212, 247
    translation 2:212
Kitab al-manazir 2:370
Kitab al-mijisti. See Almagest
Kitab al-shifa 2:391
Kitab al-zij 2:391
Kite 3:536, 567
Knitting machine 3:531, 566
Ko Yu 3:477, 566
Königsberg bridge problem 2:222, 241
Köning, Friedrich 3:523
Kosher foods 1:17
Krebs, Nicholas. *See* Nicholas of Cusa
Kulliyat 1:111

L

L'arte vetraria 3:568
L'Enfant, Pierre Charles 3:560
L'Homme, et un traité de la formation du foetus 2:237
L'Homme-machine 1:97
L'Hôpital, Guillaume de 2:234
La bilancetta 2:377
Lagrange, Joseph-Louis 2:234, 261
Lamp 3:429 (ill.)
Language theory 2:265
Languedoc Canal. *See* Canal du Midi
Lantern chimney 3:429
Lao-tzu 1:21
Laplace, Pierre-Simon 2:234, 368, 384
Larrey, Dominique-Jean 1:135
The Last Supper 3:550
Latent heat 2:361, 362, 397
Laughing gas 1:108
Lavoisier, Antoine-Laurent 2:284, 306, 351, 382–388, 383 (ill.)
Law of gravitation 2:353

Index

Laws of motion 2:385
Laws of planetary motion 2:351, 381
Laws of thermodynamics 2:362
Laws of universal gravitation 2:367
Leap years 3:510, 560
Lee, William 3:531, 566
Leeuwenhoek, Antoni van 1:11, 88, 98, **118–119**, 119 (ill.)
Lehmann, Johann Gottlob 2:285, 365, 398
Leibniz, Gottfried 2:201, 232 (ill.), **248–249**, 248 (ill.), 385, 386, 3:558
    Bernoulli, Jakob 2:255
    binary number system 2:181
    calculus 2:157, 231, 249
Leonardo da Vinci. *See* Da Vinci, Leonardo
Leonardo of Pisa. *See* Fibonacci, Leonardo
Leucippus 2:304
Levers 3:416, 471, 472–474, 474 (ill.)
    classes of 3:481
Levi ben Gerson 2:221
Leyden jar 2:364
Li Shih Chen 1:134
Li Yeh 2:166, 261
Liber abaci 2:188, 245
Liber continens 1:39
Liber quadratorum 2:246
Library of Alexandria 2:198–199
Liburnian 3:489
Light, corpuscular theory of 2:344
Light, speed of 2:358
Light, wave theory of 2:344
Lighthouse of Alexandria 3:458
Lightning rod 2:364, 3:543, 544
Lilivati 2:222
Lind, James 1:102
Lindemann, Ferdinand von 2:201

Linnaeus, Carolus 1:79, 83–85, 110, 119, 136, 140
    temperature scale 2:392
Linnaeus, Carolus 1:13, **119–121**, 120 (ill.)
Linné, Carl von. *See* Linnaeus, Carolus
Lippershey, Hans 3:564, 566
Literacy 3:513
Lithographic process 3:523, 571
Lithomists 1:53
Liu Hui 2:262
Llama 3:434 (ill.)
Llull, Ramon 2:342
Lobachevsky, Nikolai Ivanovich 2:207
Locke, John 1:99, 139, 2:351
Logarithmic tables 2:267
Logarithms 2:155, 213–216
    Napier, John 2:213, 267
Logic
    Aristotle 2:155, 177–178
    Euclid 2:204
    mathematics 2:176
Longitude 3:511, 565
Lu Pan 3:567
Lull, Ramon 3:558
Lunar calendars 2:293, 3:503
Luther, Martin 3:517, 518 (ill.)
The Lyceum 1:109

M

Ma Chün 3:567
Machines and Motion 3:**470–482**
Machines, simple 3:416
Machu Picchu 3:455 (ill.)
Magic squares 2:166, 167 (ill.)
    combinatorics 2:220
    Yang Hui 2:167
Magnetic compass 3:419, 496–497
    Ma Chün 3:567
Magnetism 2:399

Index

Maimon, Moses ben 2:334
Maimonides 2:334
Malaria 1:48
Malpighi, Marcello 1:88, 134
Ma'mun. *See* Al-Ma'mun
The Mansuri 1:43
Maricourt, Petrus Peregrinus de 2:363, 399
Marmas 1:25
Mary the Jewess 2:339, 399
Mass 2:352
Mathematical formulas 2:391
Mathematical notation
    Egypt 2:162
Mathematical symbols 2:195 (ill.), 266
Mathematical Treatise in Nine Sections 2:166
Mathematics
    Alhazen 2:370
    Alphabetic notation 2:182
    China 2:147
    Gambling 2:217
    Greece 2:149, 171–178
    India 2:168–171
    Mayan 2:169
    Mesopotamia 2:147
    symbols 2:194
Mathematics, in Greece 2:171–178
Mathematics, in India 2:168–171
Mathematics, in the Near East 2:157–162
Mathematikoi 2:253
Mather, Cotton 1:102
Matter, theories of
    early Greek theories 2:297–308
    four-elements theory 2:277, 304, 310
    particles theory 2:356
    Pythagorean theory of the universe 2:300
    single-substance theories 2:299
Maupertuis, Pierre de 2:234
Mauriceau, François 1:74
Maxwell, James Clerk 2:234, 363
Maya
    mathematics 2:169
    pyramids 3:453
    zero 2:190
Mazarin Bible 3:515
Measurements, standardization of 3:562
Mechanica 2:242, 3:472
Mechanical clocks 3:508
Mechanics 2:254
Mechanism 1:11, 52, 97
Medical astrology 2:341
Medical ethics 1:4, 30
Medical schools, rise of 1:56
Medical textbooks 1:39, 61
    midwifery 1:127
Medicine
    early laws regulating practice 1:52, 131
    Eighteenth century advancements 1:99–108
    saints, relics and pilgrimages 1:49
    superstition 1:46
    women in. *See* Women in science and technology
    women's health 1:70–78
Medicine in Ancient Egypt 1:14–19
Medicine, early Greek influence on 1:29–36
Medicine, influence of superstition on 1:42–54
Medieval Europe, Alchemy and Astrology in 2:331–342

Medieval Islam, great doctors of 1:36–42
Meditations on First Philosophy 2:237
Mediterranean Sea 3:484
Mela, Pomponius 2:399
Menaechmus 2:203
Menageries 1:83
Menkaure 3:449
Menstruation 1:71
Mental illness 1:90–91
    Pinel, Philippe 1:136
    shock therapy 1:92
Mental institutions 1:8
Mercator maps 2:229 (ill.)
Mercator, Gerardus 2:229
Mercury 3:521
Mercury barometer 2:402
Mercury thermometer 2:361, 395
Merian, Maria Sibylla 1:69
Meridians 1:20
Mersenne, Marin 2:209, 236, 244, 250, 262
Mersenne primes 2:262
Mesmer, Franz Anton 1:137
Mesmerism 1:137
Mesoamerica, development of cities in 3:452–454
Mesopotamia
    agriculture 3:412
    astrology 2:288
    canals 3:463
    cities 3:448
    mathematics 2:180
    timekeeping 2:290
    writing 3:440
Metallurgy 3:425
Meteoritics 2:393
Metius, Jacob 3:564
Metric system 2:261, 262
Michelangelo 3:567
Micrographia 1:89, 118, 2:396

Micrometer 2:380
Microscope 1:11, 88–89, 89 (ill.), 3:564
Middle Ages
    canals 3:465
    mental illness 1:90
    superstition and medicine 1:42
Middle East
    Euclid 2:206
    mathematics 2:152
    optics 2:281, 322
    physical sciences 2:280
Middle East, physical sciences in 2:320–330
Midwives 1:72, 73, 127
Milky Way 2:303 (ill.)
Mineralogy 2:390
Minus symbol 2:266
Mita of Phrygia 3:563
Mohenjo Daro 3:451, 460
Mona Lisa 1:68 (ill.), 3:550
Monads 2:249
Mondeville, Henri de 1:57, 131
Mondino dei Liucci 1:56, 66
Monge, Gaspard 2:229, 250, 258, 262
Montagu, Mary Wortley 1:102, 135
Montgolfier, Jacques-Étienne 3:532, 568
Montgolfier, Joseph-Michel 3:532, 568
Month 3:504
Moon 2:286, 290 (ill.)
Moscow Papyrus 2:162
Moses ben Maimon 2:334
Motion, laws of 1:14, 2:258, 262, 346, 352, 353
Motion, studies of 2:231
Mountains, formation of 2:391
Movable-type press 3:417, 494, 495, 512, 515, 570

Index

Müller, Johann. *See* Regiomontanus
Multiplication symbol 2:263
Mummification 1:**14–19**, 15 (ill.)
Music and mathematics 2:174, 255

N

Napier, John 2:155 (ill.), 267
    logarithms 2:213
Napier's bones 2:267
Natron 1:16
Natural history 2:392
Natural logarithms 2:215
Natural theology 1:97, 136
Nautical compasses 2:399
Navigation 2:227
Naviglio Grande 3:465
Near East
    canals 3:464
Near East, mathematics in 2:**157–162**
Needham, John Tuberville 1:107
Negative numbers 2:148, 163, 166
    Li Yeh 2:261
Neolithic Revolution 3:424, 436, 531
Neptunist theory 2:366
Neri, Antonio 3:568
New Experiments Physio-Mechanicall, Touching the Spring of the Air and its Effects 2:373
New World. *See* Exploration of new lands, impact on the sciences
Newberry, Nathaniel 3:521
Newcomen, Thomas 3:421, 528, 556, 562, 568
Newspapers 3:520–522, 522 (ill.)
Newton, Isaac 1:13, 14, 2:201, 207, 231 (ill.), 262, 284 (ill.), 296, 345 (ill.), 352–354, **385–388**, 385 (ill.)

calculus 2:157, 231
corpuscular theory of light 2:356, 386
gravity 2:283
laws of motion 2:313, 352, 353
polar coordinates 2:228
scientific revolution 2:344
Nicholas of Cusa 2:346, 375, 399
Nine Chapters of Mathematical Art 2:262
Nine Chapters on Mathematical Procedures 2:166
"95 Theses" 3:518
Nitrous oxide 1:108
Non-Euclidean geometry 2:197, 241
    Gauss, Carl Friedrich 2:258
    parallel postulate 2:207
Number Systems, early development of 2:**178–186**
Number theory 2:152, 172, 258
Numbers 2:**208–216**
    negative numbers 2:148
    irrational numbers 2:240
    lucky and unlucky 2:222
Numbers, patterns and possibilities 2:**216–223**
Numerology 2:216

O

Obstetrics. *See* Gynecology and obstetrics
    See also Midwives and Women, medical discovery of
Occupational health 1:100, 103, 104
    Paracelsus 1:122
    Pott, Percivall 1:136
Octahedron 2:319
Offfroy de La Mettrie, Julien 1:97

Index

Olivi, Peter John 2:398
Omar Khayyam 2:154, 206 (ill.), 263
   algebra 2:212
On Floating Bodies 3:537
On the Art of Hunting with Hawks 1:131
On the Equilibrium of Planes 3:537
On the Movements of the Heart and Lung 1:114
On the Nature of Things 2:304
On the Sphere and Cylinder 3:536
The Opening of the Universe 2:256
Optics
   Al-Kindi 2:398
   Albertus Magnus 2:357
   Alhazen 2:357
   Bacon, Roger 2:357
   Grosseteste, Robert 2:395
   light, theories of 2:356
   Middle East 2:281, 322
   Newton, Isaac 2:386
   rainbows 2:329
Opus majus 3:539
Opus minus 3:539
Opus tertium 3:539
Oresme, Nicole 2:347
Organic chemistry 1:95
Oscillation 2:234
Osiander, Andreas 2:349
Oughtred, William 2:213, 263, 267
Ovaries 1:131
Oxygen 2:360, 400

P

Paitammaha siddhanta 2:226
Panini 2:265
Pantocrator 1:43
Paper 3:417 (ill.), 514
   Egypt 1:18
   Su-sung 3:554
   Ts'ai Lun 3:444
   See also Papyrus
Papin, Denis 3:524, 527 (ill.), 568
Pappus 2:265
Papyrus 1:18, 2:162 (ill.), 3:446
Paracelsus 1:10, 12, 59–61, **121–123**, 121 (ill.), 2:342
Parallel postulate 2:204, 240
   non-Euclidean geometry 2:207
   Omar Khayyam 2:206
Paré, Ambroise 1:11, 48, 135
Parmenides of Elea 2:176, 277, 302
Parthenon 3:455, 456 (ill.)
Pascal, Blaise 2:156, 218, 236, **249–252**, 250 (ill.), 3:524
   hydraulic press 3:471
Pascal's triangle 2:222, 251, 256
   Yang Hui 2:268
   See also Magic squares. 2:167
Pasteur, Louis 1:13, 101 (ill.), 107, 108
Patterns and possibilities in numbers 2:216–223
Paul of Aegina 1:72, 74, 135
Pen ts'ao kang mu 1:134
Pendulum clock 2:378, 379, 3:509 (ill.)
Pennsylvania Gazette 3:543
Pentateuch 1:17
Pentaconters 3:487
Persian Royal Road 3:466
Personality types 1:33
Petit, Jean Louis 1:107
Peuerbach, Georg von 2:347, 403
Pharos Lighthouse 3:458
Philetas 1:72
Philip II of Macedon 3:570
Philolaus 2:300
Philoponus, Johannes 2:345, 392, 398
   first law of motion 2:346

Index

Philosophiae naturalis principia mathematica 1:14, 2:352, 385, 387
Phlegm 1:31
Phlogistan theory 2:358, 359
Phoenicia
    boat and shipbuilding 3:485
Phonograms 3:443
Phosphorus 2:306
Photosynthesis 1:99, 133
Physica 1:82
Physics 2:312
Physiology 1:113
π. *See* Pi. 2:148
Pi 2:148, 164, 172, 194
    Al-Kashi 2:260
    Archimedes 2:254, 3:536
    Aryabhata 2:236
    China 2:163
    Li Hui 2:262
    Tsu Ch'ung-chih 2:266
Pi Sheng 3:418, 498, 570
Piano 3:561
Pictograms 3:443
Pietá 3:567
Pilgrimages 1:50
Pineal gland 1:94
Pinel, Philippe 1:136
Pitta 1:25
Place value 2:158, 164, 168, 187
Place value system 2:161
The Plague. *See* Black Death
Plagues 3:456
Planck, Max 2:358
Plane geometry 2:151
Planetary motion, laws of 2:344, 351, 352
    Newton-Hooke controversy 2:396
Planetary orbits 2:153 (ill.), 311, 352, 381, 382
Planets 2:286

Plants
    classification 1:136
    domestication 3:434–435
    respiratory cycle of 2:360
Plastic surgery 1:25, 1:96
Plato 2:175, 198, 265
    four-elements theory of the universe 2:310
    horoscopes 2:336
Platonic solids 2:319
Plus symbol 2:194, 266
Pneumatic theory 1:34
Pneumatica 3:564
Polar coordinates 2:228
Pollaiuolo, Antonio 1:66
Polling, statistics and probability theory in 2:219
Pollio, Marcus Vitrivius 3:574
Pont du Gard 3:463 (ill.)
Ponte Vecchio 3:562
Poor Richard's Almanack 3:543
Porcelain 3:496
Posidonius 2:316, 389
Postal system 3:466
Postulates 2:204
Pott, Percivall 1:104, 136
Power 3:555
Practicae geometriae 2:246
Practice 1:28
Praying Hands 1:67
Precious Mirror of the Four Elements 2:166, 256
Preformationism 1:97–98
Pressure cooker 3:524, 527–528, 568
Pressure, Fluids and Heat 3:523–526
Prévost, Pierre 2:362
Priesthood 3:448
Priestley, Joseph 1:108, 2:284, 384, 400
Prime mover concept 2:310, 335

Index

Prime numbers 2:208-209, 238
   cryptography 2:209
   Mersenne primes 2:262
Principia philosophiae 2:237
Principia. See *Philosophiae naturalis principia mathematica*
Principle of least action 2:234
Printing. See Printing and the printing press
Printing and the printing press 1:9, 3:417, 420, 440, **512–523**, 516 (ill.), 546
   China 3:494–495
   influence on medicine 1:61
Prism 2:323 (ill.), 357
Probabilistic risk assessment (PRA) 2:219
Probability theory
   Buffon, Georges Louis Leclerc de 2:392
   Fermat, Pierre de 2:243
   Pascal, Blaise 2:250, 251
Probability theory. See Statistics and probability theory
Proclus 2:265
Prognosis 1:30
Projectile motion 2:402
Projective geometry 2:250, 258
   environmental 2:230
   Monge, Gaspard 2:262
Proof 2:172
Psychiatry 1:90
Ptolemy 2:161, 280 (ill.), **388–389**, 389 (ill.)
   cosmology 2:278
   critiques 2:283
   earth, circumference of 2:233
   geocentric theory 2:388
   geography 2:316
   influence on science 2:308–320
Public health 1:13, 55, 100, 104–105
   Black Death 1:59
   England 1:105
   Frank, Johann Peter 1:130
Pulley 3:416, 474–475
Pulmonary circulation 1:93, 132
Purging 1:32
Pyramid of Zoser 2:289 (ill.)
Pyramids
   Imhotep 3:449
   Maya 3:453
Pythagoras 2:150, 173–175, 173 (ill.), 220, **252–253**, 252 (ill.)
Pythagorean theorem 2:173, 252
   Euclid 2:240

Q

Qi 1:20
Qin Jiushao 2:256
Quadrant 2:327
Quadrivium 2:332
Quantum theory 2:358
Quarantine 1:55, 59
Quinine 1:48
Quipu 3:446

R

Rainbows, formation of 2:206, 329
   Alhazen 2:323, 372
Rational soul 1:94
Rationalism 2:237, 248
Ray, John 1:136
Réaumur, René-Antoine Ferchault 1:138
Recorde, Robert 2:194, 266
Recursive series 2:246
Redi, Francesco 1:107
Reflection 2:322
Reformation 1:61, 97–98
   printing and the printing press 3:517–519
Refraction 2:322 (ill.), 324 (ill.), 357

Index

Regiomontanus 2:227, 283, 345, 403
Relative motion 2:347
Religion and medicine 1:49
Religion and science 1:54, 2:351
    Copernicus, Nicolaus 2:375
    Galileo 2:350, 378
Religion and technology
    3:480–482
Religion in India 1:28
Renaissance 1:5, 45, 61
    biology 1:84
    surgery 1:96–97
Renaudot, Théophraste 3:523
Reproductive system in females
    1:131
Republic 2:198, 265
Retrograde motion 2:287, 316
Rhazes 1:2, 6, 12, 36, 39, 41–43,
    123–124, 123 (ill.), 132, 2:280,
    400
Rhind Papyrus 2:162, 254
Rhind, A. Henry 2:162
Ricci, Matteo 3:553
Rice 3:434
Richard II of England 1:9
Richard of Wallingford 3:570
Roads 3:416, 466
    Rome 3:467–470
Robert of Chester 2:212, 330
Rock cycle 2:396
Rockets 3:499
Rod numerals 2:187
Roman numerals 2:154, 183–186,
    184 (ill.)
Rome 3:457
    Appian Way 3:559
    boat and shipbuilding 3:489
    roads 3:467–470
    water management 3:462
Rømer, Ole 2:395
Rousseau, Jean-Jacques 2:351
Rousset, François 1:74

Royal Road of Persia 3:466
Rozier, Jean François Pilatre de
    3:532
Rubaiyat 2:263
Ruffini-Horner procedure 2:256
Rusting 2:360

S

Saint Mary of Bethlehem mental
    institution 1:8, 9 (ill.), 91
Sanguine personality 1:33
Sanitary laws 1:9, 59
Savery, Thomas 3:421, 556, 568,
    571
    steam pump 3:528
The Sceptical Chymist 2:342, 373
Scheele, Karl Wilhelm 2:360, 400
Schöffer, Peter 3:546
Scholasticism 2:334
School of Athens 1:79 (ill.)
Schools and universities. *See*
    Universities
**Science, influence of Aristotle
    and Ptolemy on 2:308–320**
Scientific method 2:284, 350
    Bacon, Roger 3:538
    Galileo 2:377
Scientific Revolution 2:342–354
    Copernicus, Nicolaus 2:284,
        344, 374
    Galileo 2:344
    Kepler, Johannes 2:380
    Newton, Isaac 2:344
    physical sciences 2:281
Screws 3:416, 472
Scribes 3:448
Scurvy 1:13, 100, 102–103
Sea Mirror of Circle Measurements
    2:166, 261
Seasons 3:503
Secant 2:226
Second law of motion 2:352

Index

Seed drill 3:574
Seed germination 1:81
Seismology 2:365
Senefelder, Aloys 3:523, 571
Seven Wonders of the World 3:456, 458
Sexagesimal system 2:158, 159, 180
    minute and hour 2:160
Shang dynasty 3:491
Sharp, Jane 1:78
Shih Shen 2:400
Ships and shipbuilding. *See* Boat and shipbuilding
Shirazi, Qutb al-Din al. *See* Shirazi, Qutb al-Din al
Shock therapy 1:92
Shuttleloom 3:531
Siddhartha, Gautama 1:23
Sidereus nuncius 2:349, 377
Sidesaddle 3:559
Sieve of Eratosthenes 2:208, 238
Silk 3:491 (ill.), 492
Silk Road 3:492
Sistine Chapel 3:567
Slavery, impact on technology 3:479–480
Slide rule 2:213
    Napier, John 2:267
    Oughtred, William 2:263
Smallpox 1:11, 13, 25, 100–102
    inoculation 1:25, 26 (ill.), 101–102
    Montagu, Mary Wortley 1:135
Smeaton, John 3:572
Smellie, William 1:78
Snel's law of refraction 2:357
Solar calendars 3:503
Solar eclipses 2:291
Solvents 2:340
Soranus of Ephesus 1:71
Spallanzani, Lazzaro 1:138
Spanish conquistadors 1:87 (ill.)

Sperm 1:98, 119
Spina, Alessandro di 3:572
Spinoza, Benedict 2:248
Splenectomy 1:96
Spontaneous generation 1:14, 107, 108, 119, 138
    Spallanzani, Lazzaro 1:138
Spring (clocks) 3:509
Square roots 2:163, 164
Squaring the circle 2:200–201, 259
Sridhara 2:170
St. Peter's Basilica 3:567
Stahl, George Ernst 2:359
Star catalogues
    Al-Sufi, Abd al-Rahman 2:401
    Almagest 2:389
    India 2:293
    Shih Shen 2:400
Stars, study of. *See* Astronomy
Static electricity 3:564
Statics 3:525, 537
Statistics and probability theory 2:216–220
Statue of Zeus at Olympia 3:458
Steam engine 2:397, 3:421, 478, 524, 527–530, 530 (ill.)
    Hero of Alexandria 3:564
    Newcomen, Thomas 3:528, 568
    Papin, Denis 3:569
    Savery, Thomas 3:571
    Watt, James 3:555
Steam pump 3:421, 528
Steno, Nicolaus 2:285
Stensen, Niels 2:365
Step Pyramid 3:449, 450 (ill.), 547
Stephanus of Calcar, Johannes 1:70
Stirrup 3:419, 438
Stone Age 3:423
Strabo 3:563
Stratigraphy 2:398
Su-sung 3:418, 498, 506, **551–553**
Submarines 3:486, 487, 562

Index

Sufi, Abd al-Rahman al. *See* Al-Sufi, Abd al-Rahman al
Sui Dynasty 3:493
Sulfuric acid 2:341, 401
The Sum of Perfection 2:401
Sumer
    development of cities 3:448
    mathematics 2:157
    wheel and axle 3:476
    writing 3:440
Sun, center of universe. *See* Heliocentric theory
Sundials 3:506–507, 507 (ill.)
Sung dynasty 3:495, 552
Supernovas 2:291, 497
Superstition and medicine during the Middle Ages 1:42–54
Surfaces. *See* Angles, Curves and Surfaces
The Surgeon Dentist 1:108, 129
Surgery 1:56
    advancements during the Renaissance 1:96–97
    Eighteenth century 1:106
    Egypt 1:3
    India 1:30
    military techniques 1:96–97, 135
Susruta Samhita 1:24, 27
Swammerdam, Jan 1:12, 88, 89, 98, 138
Syllogism 2:176, 177
Symbols, mathematics 2:194
Synagoge 2:265
Syphilis 1:10, 60, 122, 130
Syringe 2:250, 251
System of a Complete Medical Policy 1:130
Systema naturae 1:120

T

T'ang Dynasty 3:494, 552
Tabulae anatomicae sex 1:125
Tagliacozzi, Gaspare 1:97
Tahafut at-tahafut 1:111
Takakazu Seki Kowa 2:266
Tale of Two Cities, A 1:104
Tangent 2:226
Taoism 1:21
Tartaglia 2:260
Tartaglia-Cardano controversy 2:260
Tear ducts 1:40
Technology and invention in China 3:490–502
Technology, early developments in 3:421–430
Technology, impact of religion on 3:480–482
Technology, impact of slavery on 3:479–480
Telescopes 2:285, 376
    Galileo 2:349, 377
    Lippershey, Hans 3:566
Temperature scales 2:285, 360
    Celsius, Anders 2:392
    Farenheit, Daniel 2:395
Temperature zones 2:399
Temple of Artemis at Ephesus 3:458
Ten Classics of Mathematics 2:166
Teotihuacán 3:453 (ill.)
Tetrahedron 2:319
Texaurus regis Franciae 3:573
Textbooks
    geometry 2:239
    medical illustrations 1:64
    translations 2:330
Thales of Miletus 2:195, 275, 299, 402
Themistocles 3:488
Theodoric of Freiburg 1:56, 2:325, 329
Theophrastus 1:81, 138
Theorems 2:204

Index

Theories of matter. *See* Matter, theories of
Theory of animal electricity 1:91, 130
Theory of animal magnetism 1:137
Theory of embodiment 1:98
Theory of human tissue 1:136
Theory of Mathematics 2:259
Theory of the Earth 2:366
Thermodynamics 2:355, 362
Thermometer 2:360, 361 (ill.), 376, 3:562
Thermometer, mercury 2:361, 395
Thermoscope 2:360
Third law of motion 2:258
Thompson, Benjamin 2:362
Tiahuanaco 3:454
Tides, theory of 2:395
Timaeus 2:336
Timekeeping 2:289, 3:502–512
    astronomical clock 3:418, 494, 506
    China 3:418, 497–498
    Greenwich mean time (GMT) 2:367
    watches 3:509, 564
    See also Calendars, Clocks
Titan 2:378, 380
Titian 1:70
Tobacco 1:12 (ill.)
Toilets 3:560
Tomb of Mausollos 3:458
Topology 2:223, 241
Torah 1:17
Torricelli, Evangelista 2:402, 563
Torsion balance 2:364
Tourniquets 1:107
Tracheotomy 1:96
Traité de la lumière 2:380
Traité de la section perspective 2:258
Transmutation 2:340
Treatise on Calculation with the Hindu Numerals 2:247
Treatise on Smallpox and Measles, A 1:39
Treatise on the Theory and Practice of Midwifery 1:78
Trigonometry 2:151, 224–227
    Abu'l-Wefa 2:152
    Aryabhatiya 2:235
    Beg, Elugh 2:266
    India 2:226
    Middle East 2:328
    navigation 2:227
    warfare 2:228
Trireme 3:488
Trisecting an angle 2:201–202
Trivium 2:332
Trotula 1:73
Ts'ai Lun 3:417, 493, 514, 553–555
Tseng Jung-Liang 3:499, 572
Tsu Ch'ung-chih 1:266
Tull, Jethro 3:574
Turing machine 3:558
Turing, Alan Mathison 3:558
Tusi, Nasir Ad-Din. *See* Al-Tusi, Nasir Ad-Din Al
Two New Sciences 2:350, 378
Tyndall, John 1:107
Tyrtamus. *See* Theophrastus

U

Uniform acceleration 2:231
Uniformitarianism 2:398
Unireme 3:487
Universal gravitation, law of 1:14
Universe, formation of 2:368
    Chu Hsi 2:394
Universities 1:56, 2:333–334
    Ecole Polytechnique 2:230
    House of Wisdom 2:321
    Salerno 2:333

Index

University of Padua 2:334
University of Salamanca 1:73
Unsolved problems in geometry 2:**195–207**
Uranus 2:285, 354, 368, 396
The Urinal of Physick 2:266
Ussher, James 2:369

V

Vacuum pump 3:556, 571
Vagina 1:75
Vaiseshika Sutras 2:293
Varahamihira 2:295
Vardhamana 2:293
Varro, Marcus Terentius 2:331
Vata 1:25
Vedism 1:24
Velocity 2:355
The Venerable Bede 1:46
Vernacular texts 3:518
Verrocchio, Andrea del 3:550
Versuche einer Geschichte von Flotz-Gebrugen 2:398
Vesalius, Andreas 1:10, 35, 52, 69, 70, 75, **124–126**, 125 (ill.), 129, 135
Via Appia 3:469
Viète, François 2:194, 213 (ill.), 268
    algebra 2:212
Vigesimal system 2:169, 180
Vigevano, Guido de 3:573
Vision 2:323
    See also Optics
Vitalism 1:93
Vitamin C 1:13, 102
Vitruvius 3:574
Volta, Alessandro 1:91, 2:364
Voltaire 2:351

W

Wallis, John 2:379
Wantzel, Pierre 2:202, 203
Warfare
    Greek fire 3:560
    pike 3:570
    surgery 1:96–97
    tanks 3:573
    trigonometry 2:228
Washington, D.C., planning of 3:559
Water frame 3:559
Water management 3:437, 460
Waterwheel 3:477
watt 3:555
Watt, James 2:397, 3:421, 478, 529, 557 (ill.), **555–558**, 568
Wave motion 2:234
Wave theory of light 2:344, 356, 357
    Huygens, Christiaan 2:378, 380
Week 2:290
Weiditz, Hans 1:68
Werner, Abraham Gottlob 2:285, 366
Wesel, Andries van. *See* Vesalius, Andreas
Wheel and axle 3:475–477, 476 (ill.)
Wheelbarrow 3:417, 476, 493, 566
Wheels 3:477
Whetstone of Witte 2:194, 266
Whitney, Eli 3:574
Widman, Johannes
Wiles, Andrew 2:244
Wilkinson, John 3:557
Willis, Thomas 1:91, 139
Willughby, Francis 1:136
Winch 3:416
Windmill 3:477
Witch burnings 1:75 (ill.)
Witch of Agnesi 2:254
Wöhler, Friedrich 1:94

Women in science and technology 1:69, 2:399
   See also individual names:
     Agamede
     Agnesi, Maria Gaëtana
     Anne of Bohemia
     Agnodice
     Bingen, Hildegard von
     Cellier, Elizabeth
     Colinet, Marie
     Fabiola
     Fabrici, Girolamo
     Fidelis, Cassandra
     Galindo, Beatrix
     Germain, Sophie
     Herschel, Caroline
     Hypatia of Alexandria
     Mary the Jewess
     Philetas
     Trotula

Women, medical discovery of 1:70–78
Woodcuts 3:495
Wren, Christopher 1:139, 2:379, 3:528
Writing 3:439–447
   paper 3:444
   papyrus 3:446
   tools 3:442
Wu ching tsung yao 3:572
Wu Hsien 2:401
Wu Xien 2:401

X

Xenophanes 2:302
Xiguus, Dionysius 2:191

Y

Yang Hui 2:167, 222, 268
Yang Kuang 3:464
Yang Ti 3:464
Yellow bile 1:31
Yin/Yang 1:20, 22 (ill.), 2:403
Yoga 1:20

Z

Zeno of Elea 2:150, 176, 277, 302
Zeno's paradoxes 2:150, 176–177, 277
Zero 2:155, 189
   Brahmagupta 2:256
   development of 2:148, 170
   India 2:192
   Mayan 2:190
Zodiac 2:276 (ill.), 289, 339 (ill.)
Zoological gardens 1:83
Zoology 1:80
   Albergus Magnus 1:82
   Gesner, Conrad 1:130
Zou Yan 2:403
Zu Chongzhi. *See* Tsu Ch'ung-chih